SPECIAL OPERATIONS AVIATION

SPECIAL OPERATIONS AVIATION

The men and machines of the élite units

Patrick Allen

Airlife
England

Copyright © 1999 Patrick H.F. Allen

First published in the UK in 1999
by Airlife Publishing Ltd

British Library Cataloguing-in-Publication Data
A catalogue record for this book
is available from the British Library

ISBN 1 84037 013 0

Typeset by Servis Filmsetting Ltd, Manchester, England
Printed in Hong Kong

Airlife Publishing Ltd
101 Longden Road, Shrewsbury, SY3 9EB, England
E-mail: airlife@airlifebooks.com

CONTENTS

ACKNOWLEDGEMENTS

I would like to thank the following for their kind assistance in preparing this book:
Arthur Clarke DFC; Commander P.J. Craig RN (rtd); Jerry Shaw and all at the Fleet Air Arm Museum; Polizeioberrat (Lt-Col) Harald Krahmuller and all at Grenzschutzfliegerstaffel Nord; Captains Paul Hannant and Peter Ashby, Devon & Cornwall Police Air Support Unit; Marielle Winnubst and all at 298 and 300 Squadrons RNLAF/KLu; Jack Satterfield, Boeing Helicopters; HM Coastguard/Bristow Helicopters Limited, Portland; Ray Crockett of Lockheed Martin; Peter March, Barry Wheeler, David Oliver and Mike Coombes for all their kind help.

INTRODUCTION

Today, Special Operations Aviation encompasses a wide spectrum of activities both military and para-public. Over the past few years many military forces, particularly in the West, have reduced in size and consolidated their forces with the emphasis towards a more mobile and flexible strategic doctrine. Although the threat of large-scale conflict with the former Soviet Union has reduced, the threat from smaller regional conflicts has increased. This new threat is now global and many military planners have opted for aviation to provide their strategic and tactical mobility and flexibility. This in turn has brought the spotlight onto specialised military units who rely on aviation as their prime means of mobility. Many nations are developing and expanding the roles for their airborne, airmobile and amphibious/commando forces, each requiring dedicated and specialised aviation support.

This new military doctrine has also increased the role of advance/special forces whose clandestine intelligence-gathering roles require long-range covert aviation support operating within high-threat areas, often deep behind enemy lines, requiring the very best in aviation technology, survivability and airmanship skills. To these roles have been added humanitarian missions and Combat Search & Rescue (CSAR) along with combined air operations involving large numbers of fixed- and rotary-wing assets operating as a package to complete missions, and all working as one.

As the military have turned to aviation, so have many of the world's para-public organisations including law enforcement, Emergency Medical Support (EMS), fire fighting and mountain rescue etc. Like the military, para-public support has benefited from recent advances in technology including digital radio communications, night vision and electro-optics imaging systems and aviation technology. This book looks at a number of aviation units past and present who have provided specialised aviation support, and also at tomorrow's specialist aircraft including the Bell/Boeing V-22 Osprey, Boeing/Sikorsky RAH-66 Comanche and Westland/Agusta EH101 Merlin.

Schleswig-Holstein Specialeinsatzkommando squad abseiling from a GSFISt-Nord German Border Guard Bell 212. (Patrick Allen)

RAF SPECIAL DUTIES MISSIONS INTO OCCUPIED EUROPE

Almost immediately France fell, Winston Churchill directed that a secret organisation should be established to maintain contact with and provide support to resistance forces throughout France and occupied Europe. They would also deliver and pick up agents and resistance leaders and provide them with weapons, materials and specialist support. This led to the establishment of the Special Operations Executive (SOE) on 16 July 1940 with a headquarters in Baker Street, London which worked alongside the Special Intelligence Service (SIS) responsible for secret military intelligence gathering and sabotage missions throughout occupied Europe. SOE would concentrate on sabotage operations, slowing down the German war effort, supporting resistance groups, delivering weapons and explosives and preparing for the Allied invasion of Europe.

To provide the 'cloak and dagger' transport for these clandestine organisations the Royal Air Force established two permanent Special Duties squadrons which would fly covert night missions into occupied Europe providing a lifeline for resistance forces and agents. Tempsford, near Sandy in Bedfordshire, became the home of the RAF's two Special Duties squadrons, Nos 138 and 161, which moved there in the spring of 1942 and were equipped with Lysander, Hudson, Halifax and Stirling aircraft. Their role was to deliver and pick up agents, and to drop supplies of arms, ammunition, explosives, radios and a host of other military supplies to help establish and support the growing number of underground resistance groups becoming established throughout Europe.

The majority of RAF Special Duties squadron missions were flown throughout the year at low level with the pilots

The RAF Lysanders operated by No. 161 Special Duties Squadron during WW2 have become synonymous with clandestine aviation operations. (Peter March)

navigating by moonlight mainly during the two weeks before and after a full moon period. The two squadrons had two specific roles: pick-ups and parachute operations. Pick-ups were flown by the smaller Lysander and sometimes a Hudson aircraft, delivering and collecting agents and resistance leaders from within occupied Europe; parachute operations were used for the dropping of agents and supplies. Converted Halifax and Stirling bombers were used for these missions, which required the best in RAF and navigational skills with crews flying deep into Nazi-occupied Europe at night with no back-up. Missions often lasted eight hours or more and all the aircrew were hand-picked, particularly in the case of the single-pilot Lysander missions.

The Lysander flights required careful preparation, navigation and night-flying skills. Pilots needed to have at least 250 night-flying hours, which they had normally gained on night fighter or Pathfinder squadrons, before being considered eligible for Special Duties missions. For the bomber crews who volunteered for Special Duties their new role could not have been more different. Used to flying in large bomber formations at altitude on the darkest of nights, they were now required to fly low level on their own in moonlight.

SOE/SIS missions were flown throughout occupied Europe as more and more resistance forces became established in France, Holland, Belgium, Norway, Czechoslovakia and further afield in Greece and Yugoslavia. Later on in the war, Special Duties flights were flown from Brindisi in southern Italy by No. 148 (SD) Squadron with detachments based in Corsica.

By late 1943 and 1944 as the Allies prepared for the 'D' Day invasion, the work of the Special Duties squadrons increased dramatically as more and more missions were required to arm and resupply the growing numbers of resistance groups which had been so carefully trained and nurtured for this day. This extra workload saw a number of Bomber Command's squadrons temporarily assigned to supporting the two hard-pressed Special Duties squadrons at RAF Tempsford, and this included a flight from No. 149 Squadron based at nearby RAF Lakenheath, which flew Short Stirlings.

At the height of SOE operations there could often be around fifty separate missions flown during a single night throughout Europe. No. 161 Squadron Lysanders and Hudsons flew over 324 SOE missions into France during the war, and No. 138 Squadron alone dropped over 995 agents into occupied Europe. In total between April 1942 and May 1945 over 29,000 containers and 10,000 smaller packages were delivered by the Tempsford squadrons. Flight Lieutenant Arthur Clarke DFC, a Stirling navigator on No. 149 Squadron (Special Duties) Flight, recalled a typical SOE mission flown into France:

> Our Stirling had a crew of seven: pilot, navigator, bomb-aimer/front gunner, wireless operator, engineer, mid-upper gunner and rear gunner. We usually dropped two coffin-shaped wooden containers by parachute to a triangle of three torch lights at a given time. No circling was allowed; the navigation had to be precise and if the arrival was one minute out there would be no small red lights and the flight would be counted a failure. On

arrival, we would flash a recognition signal, receive a counter signal, climb a few hundred feet up to five hundred feet for the parachutes to open and then continue in a straight line for several miles before turning and flying a different route home at heights of fifty or one hundred feet. SOE aircraft carried an assortment of cargo including pigeons, radios, stores, explosives, banknotes, printing presses, spies and underground members.

On 7 March 1944, flying in Stirling 'Q' LK382, we set out on a moonlit night to a group of Maquisards in the Alpine mountain region of Haute Savoie with its deep valleys, close to Mont Blanc at 15,782 feet. Meticulous navigation preparation was vital, a flight plan being completed using Met. forecast winds and a thorough study of the route made in order to anticipate special ground features, high ground and, especially, water in the form of rivers, lakes etc.

The flight of eight hours and fifteen minutes, hedge-hopping as usual to Annecy, south of Geneva, went well. Flying low, rivers, small towns and hamlets rushed by in an eerie moonlit blur. Low-level map-reading, difficult by day, was essential by night, even when cloud sometimes obscured the moonlight. Speed was necessary in identifying features seen on the ground with those on the map. Normal dead-reckoning navigation was used to provide the basic route framework to the dropping area, when map-reading took over. Always alert, position and timing had to be constantly checked, as complacency was fatal. The main difficulty was timing, rather than track keeping.

Timing was vital and stringent conditions were applied with no second attempt being allowed. Concentration and anticipation were essential to rendezvous in a small clearing in woodland, in a deep valley or a tiny isolated field at a given time to the minute. Our navigational methods would be considered primitive today. All the crew would be alerted a minute or two before a river or water feature was reached, as this could often be seen, even through cloud. Increasing airspeed to gain time needed through wind changes, or dog-legging and reducing airspeed to lose time, was often necessary. One danger encountered at rivers was that balloons could be found there; our aircraft had cutters on wing leading edges which were supposed to cut through cables, but they later had explosives attached to them. The weather over Europe that winter was atrocious and losses were high, thought to be due to icing, in addition to the usual hazards of storms, dense low cloud, light flak and fog.

Should arrival be one minute out there would be no small red lights and the flight would be counted as a failure, as the reception committee put their lights on as soon as they heard aircraft engines, providing the timing was right. However, on this mission all went well. On arrival we flashed a recognition signal, received a counter signal, climbed a few hundred feet to allow the parachutes to open, then dropped and continued as

> briefed in a straight line for several miles, before flying another route home at 50/100 feet. During the drop a brave soul, one of the reception committee, flashed V for Victory at us with his Aldis lamp, seven hundred miles from home.
>
> Relations between aircrew and the Resistance, based on mutual respect, were excellent. At the drop of supplies, the wireless operator always dropped a container marked 'Gift of RAF Aircrew', holding news-letters, banknotes, cigarettes, coffee, chocolates, etc., which brought tremendous goodwill at the shareout.

Listed below are some of the agents into whose territory supplies were dropped by the Special Duties Flight under Squadron Leader A. Reece of No. 149 Squadron during the first three months of 1944.

6/1/1944: Stirling 'P' EH904, St Jean de Luz (Biarritz), nine hours fifty minutes.
Agent George Starr (codename 'Hilaire'), Controller of the Scientist and Wheelwright Circuits, who also later led a thousand Maquisards in the Toulouse–Avignon area. A quiet, tough, unobtrusive mining engineer who attracted followers in a quiet sort of way. His radio operator was 'Annette' (Yvonne Comeau) whose husband, a Frenchman born in England, had been killed in action in the early days of the war. Hilaire had requested an elderly man because younger Frenchmen were being picked up and sent to labour camps in Germany. However, the BBC message 'Jacqueline has a red and green coat' meant otherwise. When Annette arrived at his Pyrenean hide-out he recognised her from cricket matches in pre-war Belgium.

They travelled as brother and sister, using safe houses up to fifty miles apart, including an isolated house in the Pyrenees. Coded messages received, such as the co-ordinates of dropping zones, were memorised, decoded messages, written in shorthand, were never kept more than a few hours. They narrowly escaped betrayal at the outset by Milice, anti-Resistance French who worked with the Gestapo.

Supplies from Spain had to be shipped around the Spanish–French coastal border, where the railway tracks differed. Resistance members and escaping RAF crossed the Pyrenees travelling high and in snowy conditions at night to cover their tracks, as roads and passes were controlled by Germans with dogs. It was an arduous, eighteen-hour forced march. This particular flight caused an international incident as Spain claimed that the RAF had violated its airspace to make the drop.

10/2/1944: Stirling 'Q' LK382 South Central France; then Bordeaux, seven hours fifty minutes.
Agent Father Felix Kir, Mayor of Dijon and leader of the Resistance in Lyon. He set up an escape route for RAF aircrew who had successfully baled out. He gave them his home-made drink of white wine and Dijon blackcurrants. This was

named after him, in his honour, and Kir Royale is now a well-known aperitif.

11/2/1944: Stirling 'Q' LK382 Toulouse, seven hours thirty-five minutes.
Francis Caamaerts, codenamed 'Roger', Controller of Provence. A teacher at Penge School and a conscientious objector of great integrity, he decided that evil could not be allowed to prevail unchallenged and unchecked. His father was a Belgian man of letters and his mother a Shakespearean actress.

15/2/1944: Stirling 'J' LK504, Geneva–Annecy, six hours fifty-five minutes.
Harry Ree, Controller of the Jura Foothills and Swiss Lake Annecy area. Once shot in a lung with four other bullets in his arm, he swam a fast-flowing river and crawled four miles through woods and fields to a doctor who thought the feat barely possible.

2/3/1944: Stirling 'Q' LK382, Paris, four hours forty minutes.
François Mitterrand, Controller of Paris and post-war President of France.

7/3/1944: Stirling 'Q' LK382, Haute Savoie, eight hours fifteen minutes.
George Millar, Controller of Haute Savoie, south and east of Geneva, codenamed 'The Horned Pigeon' and noted for his success at railway sabotage.

10/3/1944: Stirling 'Q' LK382, Mediterranean coast, eight hours five minutes.
Peter Churchill, Controller with a base at Cannes, who married his courier 'Odette' who survived torture at Ravensbruck concentration camp, and Robert Boiteux, Controller of the Gardiner Circuit of Marseille, using the zoo as his headquarters. Betrayed and tortured in Fresnes Prison and Buckenwald, his suicide tablet was in a signet ring taken by the Gestapo. He survived by managing to exchange identity with an inmate who died of typhus. Later he was dropped behind the Japanese lines in Burma.

13/3/1944: Stirling 'P' LK526, Ancey/Swiss border, five hours thirty minutes.
Tommy Yeo-Thomas, a director of the French fashion house of Molyneux, codenamed 'The White Rabbit', Controller of the Swiss and Italian frontiers.

Thanks to these agents and the aircrew who supported them the Special Operations Executive proved extremely effective. Maurice Buckmaster ran the SOE (French section) with 480 agents – 440 men and forty women – of whom 120 were lost during the war, and many died in the years immediately following the war. Hitler is reported as saying that after the defeat of Britain he would be undecided whether to hang Churchill or Buckmaster first.

BORNEO JUNGLE OPERATIONS 'THE JUNGLIES'

On 2 March 1997 a Royal Navy Amphibious Task Group led by the assault ship HMS *Fearless* arrived on the coast of Brunei on the island of Borneo to disembark 800 Royal Marines from 40 Commando Group plus their RN Commando support helicopters to begin two months of jungle training and joint exercises with the Royal Brunei Armed Forces. The deployment culminated in a major joint and combined amphibious/jungle exercise known as Exercise Setia Kawan (Loyal Friends) which took place between 5 and 15 April.

The exercises were part of the Royal Navy's eight-month Carrier and Amphibious Task Group 'Ocean Wave 97' deployment to the Asian Pacific region. The deployment was intended to demonstrate the United Kingdom's continuing commitment to the region, and the Royal Navy's ability to deploy an operationally effective and self-sustaining maritime and amphibious force east of Suez for prolonged periods. It also helped to prove many of the maritime and amphibious

operational concepts destined for use within the newly formed Joint Rapid Deployment Force (JRDF).

Exercise Setia Kawan also saw the return of the Royal Navy Commando helicopter squadrons to the region after an absence of thirty-one years, working alongside Royal Marines and Royal Brunei Air Force helicopters from Nos 1 and 2 Squadrons. It was the work of the RN Commando squadrons in the jungles of Brunei, Sabah and Sarawak during the Indonesian confrontation between 1962 and 1966 that earned the squadron its nickname 'The Junglies'.

Today, there remains a small British garrison located at Seria, the centre of Brunei's oil support facilities located on the west coast, with a resident Gurkha battalion (2 Royal Gurkha Rifles). This is also the location of the Training Team Brunei, a Jungle Training School and the last level of expertise in jungle warfare within the British Army, including Special Forces. Supporting these two resident units is a flight of Army Air Corps helicopters. This presently is No. 7 Flight AAC which

An RAF No. 66 Squadron Belvedere HC Mk 1 landing at a grass airstrip in Borneo in the mid-1960s. (John Allen)

A Royal Navy Commando Whirlwind HAS7 seen at a forward base in Borneo in the mid-1960s. (John Allen)

operates three leased Bell 212s providing jungle casevac and helicopter support for both jungle training and the Gurkha battalion. The flight is highly proficient in jungle helicopter support operations including Night Vision Goggle (NVG) missions. Flown and operated by the AAC, the three Bell 212s have been leased from Bristow Helicopters Limited and arrived in Brunei to replace the ageing Scout helicopters in early 1995. The Bell 212s are ideal for jungle support missions and can carry eight to ten troops. Fitted with full NVG cockpit lighting, the Bells are equipped with a winch with 250 feet of cable and carry a stretcher for the casevac role. Flown as single pilot with a crewman trained to have some time on the controls during an emergency, the flight has four pilots and four aircrew and is regularly called upon to winch troops and casualties out of the jungle at heights in excess of 200 feet.

Indonesian confrontation and Operation Claret

Indonesia gained her independence from the Netherlands in 1949. Her new leader, President Sukarno, had a vision of forming an Indonesian Federation comprising Malaysia, Singapore and Borneo, which included the Sultanate of Brunei. This was in direct opposition to the proposed Federation of Malaysia comprising Malaysia, Singapore, Sabah, Sarawak and Brunei. President Sukarno saw this as a

threat to his ambitions of Indonesian expansion throughout the region. The 1,000-mile land border between Indonesian Kalimantan and Sarawak, Sabah and Brunei was extremely vulnerable to Indonesian insurgents.

The majority of Borneo and the border region comprises dense primary jungle with poor, almost non-existent, communications which were and still are limited to only a few rivers. An Indonesian-inspired revolt led by a young Brunei sheikh, who wanted to encompass Brunei within Indonesian Sarawak and Sabah under the name North Kalimantan, eventually led to the December 1962 Brunei Revolt. This resulted in the arrival of British forces in the region and the revolt was crushed, although over 1,000 rebels fled into the jungle to form the Indonesian-backed Clandestine Communist Organisation (CCO). Sukarno began to infiltrate Indonesian troops and communist insurgents into Sarawak, Sabah and Brunei across the land border with Kalimantan, who then worked alongside the CCO.

The Royal Navy Commando helicopter squadrons first arrived in Brunei and Borneo in December 1962 to help quell the revolt. On 1 August 1962 the Commando carrier HMS *Albion* was commissioned with both 845 NAS, with twelve Wessex HAS 1s, and 846 NAS, with six Whirlwind HAS 7s, embarked. In November she sailed for the Far East, and while in the Indian Ocean was ordered to make all speed for

Singapore to help quell the rebellion in Brunei. HMS *Albion* brought troops from Singapore to Brunei where the Commando helicopters were immediately committed to supporting troop operations ashore. Their helicopter support enabled troops to patrol vast areas of jungle, operating both from the ship and from primitive forward bases in Sarawak and Sabah. Some of these forward operating bases, such as Nanga Gaat and Belaga in Sarawak, were no more than jungle clearings with the most basic of aviation facilities. One of their priority missions was to support and resupply troops operating along the 1,000-mile border with Kalimantan. A chain of jungle camps and helicopter platforms was established along the border and Gurkha and SAS troops were deployed to act as a screen. The Commando helicopters provided their only means of mobility, and once initial contact was made with the enemy they would then fly in reinforcements. The helicopters were the troops' only means of mobility and were often deployed to remote areas for weeks at a time.

During these early years, Borneo helicopter operations demanded the highest of flying standards. Most of the interior during this period was unmapped and the Navy pilots had to make their own maps as they went along. This was no mean feat considering the lack of navigational aids aboard the Whirlwind and Wessex, and the featureless jungle terrain. In these hot tropical conditions, with the helicopters often carrying heavy loads of troops (the Whirlwind could carry four to six troops) or cargo, engine power margins were small, making flying into and out of remote jungle clearings an interesting experience. Many of the jungle sites were surrounded by tall trees which restricted the helicopters' approach and departure routes and resulted in steep descents or departures necessitating full engine power. The jungle was no place for a 'downbird', and for the Whirlwinds in particular, the use of the Leonides piston engine, which had in the past suffered a series of engine-associated failures, resulted in a number of accidents and several helicopters were lost. Other hazards included helicopters rolling off the timber platforms into ravines and tree stumps being pushed up through the aircraft floor, plus the inevitable blade tip strikes.

In 1965 HMS *Albion*, with 848 NAS equipped with their brand-new Commando Wessex Mk 5s, arrived off Brunei to relieve 845 and 846 Squadrons. 848 NAS deployed into Brunei with two flights of four aircraft each. Operating from Sibu and Nanga Gaat in support of counter-terrorist operations, they performed tasks as varied as ferrying troops and flying sick native men and women to hospital. In September 1965, 848 moved to new sites in Labuan, a small island off Brunei in the South China Sea, with a detachment at Bario 100 miles inland. It was from these two sites that the squadron continued its task of supporting troop operations along the border.

It was at this period that 848 NAS became involved in supporting the top-secret SAS Operation Claret missions flying SAS/Special Forces troops to remote border jungle landing sites to undertake cross-border surveillance missions, search and destroy raids inside Kalimantan on Indonesian forward staging posts and army camps, and ambushes on Indonesian insurgents infiltrating, or returning from missions inside, Sabah and Sarawak. The Special Boat Service (SBS)

An 846 NAS Commando Sea King flying in a Rapier missile system to a jungle clearing in Brunei during Ex Setia Kawan. A jungle fire burns in the background. (Patrick Allen)

were also keeping busy undertaking similar missions along the many rivers and islands which were also used as a means of infiltration.

The Claret missions often involved 845/846/848 Squadrons operating out of remote jungle landing sites such as Long Jawi in Sarawak. It was from these sites that the 'Junglies' flew A and D Squadrons, 22 SAS to landing platforms located deep in the jungle. The Claret operations began in June 1964 and included a number of cross-border reconnaissance, ambush and search and destroy missions up to ten miles inside Kalimantan. These raiding parties also included around forty SAS-trained indigenous tribesmen (Ibans), known as the Cross-Border Scouts, who were used for scouting and intelligence gathering. These Special Forces missions proved highly successful and the troops involved became known as the 'Tiptoe Boys'. They inflicted a heavy price on the Indonesian insurgents.

In late March 1966 848 NAS and 110 Squadron RAF flew SAS troops on the last Operation Claret mission to destroy a large Indonesian army camp inside Kalimantan. Having deployed the troops, the mission was quickly re-called as news of an army coup in Indonesia against President Sukarno became known and there was now the possibility of an end to the confrontation. A peace treaty was signed in the summer of 1966 ending hostilities, and 848 NAS handed over its task in August 1966 and returned to the United Kingdom aboard HMS *Albion*, ending the Junglies' four years of operational missions in Borneo.

The Commanding Officer of 848 NAS, who flew on the last Operation Claret mission which began on 24 March 1966, provided this first-hand account:

In early March 1966, I was one of the passengers on board a Comet 4B of RAF Transport Command flying from RAF Lyneham to Singapore to take over command of 848 Naval Air Squadron who were equipped with twenty-two twin-engined Wessex Mk 5 helicopters and were involved in the Borneo campaign. It took nearly six days to make the long and tedious journey via Cyprus, Bahrain and Gan. Normally based on board the helicopter carrier HMS *Albion*, the squadron was disembarked at the Royal Naval Air Station Sembawang on Singapore Island with one detachment operating from Labuan Island off the coast of Brunei and another from the airstrip at Bario some 100 miles inland in Borneo.

After arriving at Singapore and settling in and being briefed on the current situation it was time to visit the Borneo detachments, and I was flown to Labuan on 23 March. That evening, over a plate of tiger prawns and a beer or two, the senior pilot and I discussed the operational details. We had just about covered everything when a tasking signal arrived instructing the detachment to provide two Wessex for a troop lift the following day at Sibu, some 300 miles to the south-west. Sibu was one of the two towns of any size in Sarawak, the other being Kuching, the capital. It was situated in a flat mangrove swamp delta where the mighty Rajang river flowed into the sea, and was an HQ base for the area with an airfield suitable for regional airlines. As I read the signal I wondered why the Wessex were required at all – Sibu was RAF helicopter territory where No. 110 Squadron operated a number of Whirlwind Mk 10s. However, it sounded simple enough and an ideal chance for me to get back into harness again, so I told the senior pilot that I would grasp the opportunity.

At 0800 hours on 24 March we fired up the two Wessex, each with two pilots and an aircrewman, and taxied out and took off along the main runway of Labuan airfield in bright sunshine; vertical take-offs were avoided if possible in order to reduce mechanical wear and tear. On board we had a small maintenance 'downbird' team and some 'come in handy' spares because we were going to be a long way away from base.

The first leg was flown inland at 1,000 ft but we soon ran into heavy tropical rain which reduced visibility to about a mile. The maps we had not only lacked detail but were notoriously inaccurate. There were often huge areas of white paper with even larger features like rivers

fading into vague dotted lines. Mountain ranges were superficially shown by 'hatching' and one mountain range was known to be ten to fifteen miles out of position. As we were unsure of our exact position, I decided to turn north and head for the coast. Navigation equipment aboard the Wessex Mk 5 was non-existent. They did, however, have good radios – the ARC-52 UHF with an infinite number of dialable frequencies plus an excellent SSB HF radio. The only other device aboard the Wessex was 'Violet Picture' which was a radio homer giving a left/right indication towards a ground transmitted signal. Navigation was mostly a case of marking down unusual features on the map for future reference!

Once at the coast we were soon flying at 200 ft along the beach but in even heavier rain. The windscreen wipers, pretty useless at the best of times, were completely ineffectual, but by opening the pilot's sliding window and by flying slightly sideways at eighty knots we were able to see ahead for about half a mile. I had hoped that the weather would clear but it was still pouring down when we reached the grass airstrip at Bintulu where we planned to refuel from the dump of forty-gallon drums. We topped up using hand-operated pumps, a slow process but the only one available. We kept one engine running while doing so.

Airborne again, we turned on to a southerly heading for Sibu. About twenty-five miles away we at last broke out into clear weather and landed after what had been a very unpleasant flight lasting over four hours. Keen to hear what our task would be I was somewhat surprised to be told that it was to take fifteen SAS troops some 150 miles eastwards to Long Jawi, a remote longhouse deep in the interior about twenty-five miles from the border with Indonesia – hardly the simple troop lift in the Sibu area I was expecting! Because this was unknown territory to us we were instructed to land *en route* at Nanga Gaat to pick up a pilot from the detachment of RAF helicopters based there. Leaving most of our 'downbird' team at Sibu with the majority of stores and splitting the fifteen SAS troopers between the two Wessex we got airborne and headed east along the Rajang passing the riverside settlements of Song and Kapit to land at Nanga Gaat just over an hour later.

Nanga Gaat was a former helicopter base that had originally been established by 845 NAS in late 1963. It was situated on a promontory at an important river junction as far forward as possible to cover the border while still maintaining access to the vital river supply route; it was also home of the Temenggong Jugah, Paramount Chief of the Ibans and therefore politically important. The personnel of 845 NAS, with the help of local labour, developed Nanga Gaat into a first-class base with living quarters, showers (cold in the morning, hot in the afternoon), electric lighting, galleys, even an 'Anchor Inn' plus maintenance facilities and six landing pads for the Wessex Mk 1 helicopters. By 1966 845 NAS had returned to the UK and the base was occupied by a detachment of RAF, 110 Squadron equipped with Whirlwind Mk 10 helicopters.

On landing at Nanga Gaat there was a setback when my wingman reported that he was shutting down because of a tail rotor control problem. With the RAF pilot up front in the left-hand seat and ten of the SAS aboard, who turned out to be from the New Zealand SAS Rangers, I set off for Long Jawi but soon ran into thick cloud. Our route was leading towards the Hose Mountains and I was none too happy in view of the absence of any navigational aids. In addition, my RAF guide was beginning to look distinctly nervous, so there was nothing for it but to turn back rather than press on into the unknown.

That evening I asked the RAF detachment Operations Officer for more details of what was afoot, but didn't get very far: 'All I can tell you is that your two Wessex and two of our Whirlwinds are to do a troop lift tomorrow, but I am not allowed to tell you what it's about.' That was all I could get out of him. It all seemed far too secretive – we were, after all, on the same side!

Early next morning, with the tail rotor problem on the other Wessex rectified (or so I thought), we got airborne with the New Zealand SAS Rangers on board and followed the two Whirlwinds to Long Jawi where we landed some thirty minutes later after a flight in clear weather. Long Jawi was a longhouse on the banks of a major tributary of the Rajang river and had been the scene of one of the earliest raids by Indonesian troops during which five defenders had been killed and seven murdered in cold blood after surrendering. Retribution by 1/2 Gurkha Rifles, supported by helicopters of 845 NAS, was swift and few of the raiders survived to reach safety across the border. Since then, Long Jawi had been reinforced and now had a garrison of company strength. The task was finally revealed at the pre-mission briefing at Long Jawi. We were to lift a total of eighty-eight troops drawn from C Company, 1st Battalion King's Own Scottish Borderers and New Zealand SAS Rangers to a log platform landing site cut out of trees on a 2,000-foot-high ridge about thirty miles away on the border. Their aim was to attack an Indonesian army camp some ten miles over the border at Long Nawang.

Since early 1965 such raids, which went under the codename of 'Operation Claret', had been carried out under conditions of maximum secrecy. Every raid had to be personally authorised by the Director of Operations and their aim was to prevent the Indonesians from launching an attack. They were never mounted in retaliation and civilian casualties had to be avoided at all costs. In addition, each raid had to be self-supporting because no air support was authorised except in dire emergency. Months of intelligence gathering by covert patrols and meticulous planning lay behind these raids. Towards the end of the briefing I was told that if the raid was rumbled, or they had to make a fighting withdrawal, we would be required to haul them out

from a prearranged rendezvous site inside Indonesia. Although the chosen RV – a padang, or 'village green' – was satisfactory from a pilot's point of view, I felt some surprise that a senior and experienced member of the Labuan detachment had not been brought in on the planning of the raid at an early stage. If he had, we would have made sure that both Wessex were fitted with a door-mounted general-purpose machine-gun and that the aircrew wore flak jackets, but it was too late now. In addition, he would no doubt have recommended that three Wessex, rather than two, were more appropriate for the task; this additional level of support would have been more than welcome, as you will see.

The troops were armed to the teeth, with spare belts of ammunition draped round their shoulders, and carried heavy backpacks containing everything they might need for what would be a difficult approach through the jungle to their objective. Bare minimum fuel loads for the round trip to the landing site were calculated, and the spare lifting capacity divided by the weight of each man gave the numbers that could be carried on each sortie. Hot, high and humid conditions affect a helicopter's performance and these factors had to be taken into account as well.

Unfortunately, my wingman had had a recurrence of the tail rotor control problem on landing at Long Jawi which caused the tail to 'wag' from side to side without warning. This fault was too serious to allow him to take part in the lift – it was far more important to rectify it and have him available for extracting the raiding party if an emergency arose. As far as the lift was concerned it was going to take six hours instead of the three and a half hours we had planned on; with each Whirlwind able to carry four troops and a Wessex eight, only sixteen could be lifted on each round trip instead of twenty-four.

Guided by one of the two Whirlwinds I set off at 1030 hours towards the border with my eight troops embarked. There could be no sizing up of the landing site by circling around it, – it was a case of flying at low level over the jungle then climbing up the ridge and forward to hover over the log platform, which turned out to be only just big enough. On touchdown I kept my Wessex almost, but not quite, airborne and gradually lowered the collective lever that controlled the lift of the rotor blades to compensate for the reduced weight as the troops leapt out. With no way forward because of the trees it was a case of backing out then doing a descending right-hand turn down the side of the ridge and heading back for the next load as quickly as I could. Each time I landed back at Long Jawi I was given a rotors-running top-up of fuel while the next stick of troops embarked, and so it went on all day, shuttling back and forth, being refuelled, fed with the occasional sandwich and gratefully downing tins of fizzy drink at every opportunity – it was very sweaty work. Just over seven hours after taking off that morning, lift completed, I landed back at Nanga Gaat after an exhausting but satisfying day's flying.

The next two days were spent standing by at Long Jawi in case of an emergency call from the raiders while the snag team replaced an auto-pilot servo motor that had been the source of all the tail rotor problems in the other Wessex. Late on the second day I was told that the raid had been re-called. Apparently President Sukarno had at last, after almost four years, recognised that his efforts to prevent the formation of the Malaysian Federation had failed; political moves were afoot to bring an end to 'confrontation' and this was not the time to rock the boat.

It took the raiders two days to retract their steps but on 29 March they had got back to their original landing site. With both Wessex now available it didn't take long to lift them all back to Long Jawi. I was not looking forward to the long 450-mile flight back to Labuan via Nanga Gaat, Sibu and Bintulu so, after studying the map, I decided we could take a more direct route: we would fly north-east along the border to Bario to refuel and then home to base.

The scenery was magnificent as we flew at 1,000 ft over the long stretch of apparently uninhabited jungle in perfect weather; hills and mountains, deep river gorges with white water rushing over rapids, extraordinary outcrops of limestone draped with vegetation rising like fingers out of the rainforest, and even bright-coloured birds fluttering amongst the treetops. About halfway there we raised Bario on the radio and, after a quick refuelling, we were on our way again over the 7,000-feet-high Gunung Murud ridge to the north to land at Labuan having logged six and a half hours in the air for the day, and, as far as I was concerned, a total of twenty hours for the whole operation.

The original tasking signal had given no hint of what was in store. Nevertheless I was pleased to have played a small part in what was almost certainly the last raid in the Operation Claret series. Although it had been a bit more than I had bargained for, it had certainly got me back into the swing of things a lot quicker than I had expected!

Exercise Setia Kawan

On 2 March 1997 Nos 845, 846 and 847 NAS returned to Brunei and Borneo when the Royal Navy Amphibious Task Group, led by the assault ship HMS *Fearless*, arrived off the coast of Borneo as part of the Royal Navy's 'Ocean Wave 97' deployment to the Far East to disembark 800 Royal Marines from 40 Commando Group plus their RN Commando Support helicopters to begin two months of jungle and cross training with the Royal Brunei Armed Forces, culminating in Exercise Setia Kawan.

The exercise initially involved the deployment of Special Forces by helicopter to undertake intelligence-gathering

Commando Sea King deploying a patrol from 40 Commando RM to a jungle clearing in Brunei. The patrol had been in the jungle for several weeks. (Patrick Allen)

The US Army Special Forces developed the STABO system to haul their troops out of the jungle during the Vietnam War. A US Army Black Hawk is seen hauling US airborne troopers. (Patrick Allen)

from the door of a Sea King or, in the case of the Chinook, off the ramp, jumping into the sea with their equipment and boats. The Chinook can also undertake water landings where troops can be pushed in their boats directly into the sea and, more importantly, can be recovered by the same method, boats and all. For the BPG and other advance forces who need to be inserted inland, they can be parachuted into the jungle by C-130s or other transport aircraft. This method was perfected during SAS operations in Malaysia and Borneo in the 1950s and 1960s. The casualty rate in those early days was high and troopers were equipped with a rope by means of which they could let themselves down through the jungle canopy. Today, the preferred method is by helicopter. Troops can be winched down through the canopy and, more importantly, winched out. For helicopters without a winch, or when speed is a priority, troops can fast-rope or abseil down through the jungle canopy, although recovery by this method is usually not possible.

During the Vietnam War, American Special Forces developed the Stabilised Tactical Airborne Body Operation, known as STABO, to haul their Special Forces people out of the jungle when an urgent extraction was required, there was no suitable jungle clearing, and using one of their 'daisy cutters', a huge air-dropped bomb designed to blow down trees etc. and make an instant helicopter landing site, was out of the question. The STABO method involves the helicopter hovering above the jungle canopy and dropping down 200 feet of line with harnesses attached. The Special Forces troops then strap into their harnesses and the helicopter just pulls them out and transports them to a safe area where the helicopter can land and the troops can climb aboard. This technique was only used in dire emergencies as there are many pitfalls in swinging below a helicopter above the jungle at night travelling at 100 knots. The major disadvantage is the pilot not being aware of rising ground and increasing his height enough, or smashing the troopers into obstacles during departures and arrivals etc. Not a method for the faint-hearted, or a forgetful pilot.

The main assault for Exercise Setia Kawan was launched at 0100 hours on 6 April with a combined amphibious and heliborne assault into the Daerah Belait region with two companies of Royal Marines and RBAF being landed onto Red Beach by landing craft from HMS *Fearless*, while the six Sea Kings and RBAF Bell 212s undertook night heliborne assaults ten miles inland moving over 800 troops in two waves to take and hold the key road and river crossings at Pukit Buan over the Sangai Belait river. The exercise was an ideal opportunity to conduct a freeplay exercise with all participating forces unsighted on enemy activity, the enemy (the resident Gurkha battalion) only being constrained within time and space and not in tactical manoeuvre. This allowed all participating forces the opportunity to practise real-time battle procedures and command estimates, and to conduct operations within realistic timeframes and distances.

The night combined amphibious and heliborne assault proved highly effective and cut off the enemy force's only means of retreat away from the coastal region and into the

missions on enemy troop dispositions with the SBS covertly deploying by Sea King helicopter deep within the mangrove swamp areas of the Sungei Tutong to make their way to the proposed landing areas to reconnoitre suitable beaches for the main amphibious landing. They were followed by the Brigade Patrol Group (BPG) which conducted medium-range reconnaissance tasks ahead of the main force. Advance forces can be covertly inserted by various methods, often several weeks ahead of the main force. Teams can be inserted by submarine and small boats, or more often by long-range helicopter insertion. The Sea King and Chinook both provide a long-range night capability and teams can be heli-casted jungle interior. The heli assault proved so effective in trapping the enemy that the exercise had to be stopped for a period to allow the enemy to regroup and break out into the jungle to begin their insurgency operations deep in the interior, operating throughout the remote Hutan Simpan Labi region.

For the Commando Sea King squadrons the exercise provided an ideal opportunity to relearn jungle warfare skills, to conduct air mobility and air assault procedures in a jungle environment, and to cross train with the RBAF. The squadrons undertook NVG operations from both the ships and in the jungle. Operating in this hot environment the Sea Kings were roled to carry sixteen troops with around one hour's fuel. The pre-positioning of fuel was an important factor in planning for jungle missions with much of the tasking involving the move-

RM Commandos roping from an 846 NAS Sea King in Brunei. (Patrick Allen)

Commando Sea King pilot flying (LHS) with a Royal Brunei AF Bell 212 during the amphibious phase of Ex Setia Kawan. (Patrick Allen)

ment of troops and patrols to jungle landing sites deep within the Labi region. Missions also included day and night resupply of water and food and providing jungle casevac day and night. GPS systems cannot operate under the jungle canopy so jungle patrols used the age-old method of counting paces to estimate position, and used marker panels placed in small clearings or inflated a balloon above the canopy to mark the position for helicopters.

During the exercise an 846 NAS Sea King was called out on a jungle casevac mission with a report that a marine with a jungle patrol from Charlie Company, 40 Commando Royal Marines had suffered a snakebite and needed to be urgently airlifted out for medical attention. The patrol proceeded to clear a winch-hole in the jungle canopy for the casualty evacuation. A Sea King about to launch on a Rapier missile lift was re-tasked for the casualty evacuation, and within a short period arrived at the patrol's position guided in by an 847 NAS Lynx which was already operating in the area but carried no winch. The Sea King positioned itself just above the jungle canopy about 150 feet above the ground and the aircrewman was winched down into the winch-hole. The patrol had found the nearest hole in the canopy and then proceeded to cut away at the foliage to create the small opening for the transfer, marking the position on the ground with coloured marker panels.

Unfortunately, as the aircrewman on the ground was preparing the casualty for evacuation the helicopter's downwash toppled a 200-foot-high tree which fell onto members of the patrol waiting on the edge of the winch-hole, injuring four, one seriously. The Sea King aircrewman, who was at the time on the ground attending the snakebite casualty, was lucky not to have been hit by the tree, or worse, the falling tree snagging and pulling the winch cable as it fell. The noise of the helicopter prevented anyone on the ground from

hearing the tree as it fell, and its arrival caught the marines offguard. The aircrewman quickly winched up the casualty. The marine, in fact, had been stung by a hornet and had suffered an allergic reaction which left him in a state of shock. He was winched aboard and the aircrewman returned to collect the most seriously injured tree-fall casualty, and both were rushed to the nearest medical centre, which was at brigade headquarters in Seria, about fifteen minutes away by helicopter. As soon as there had been an assessment of the numbers of casualties on the ground a quick radio call was made to alert other helicopters to take the less seriously injured to hospital. On reaching the medical centre at Seria the casualties were assessed and a decision was made to take the marine who had suffered the most serious injuries straight to the main hospital at Bandar Seri Begawan. Unfortunately this marine suffered numerous broken bones and a broken back and was left paralysed. The incident certainly highlighted the dangers of operating in a jungle environment for both the troops on the ground and the aviators who support them.

The exercise and deployment proved a great success with both 40 and 45 Commando and Commando Aviation, all becoming fully trained in jungle warfare and amphibious operation. Within the newly formed Joint Rapid Deployment Force the exercise has helped to show the Royal Navy's ability to deploy an operationally effective and self-sustaining maritime and amphibious force east of Suez for a prolonged period and has helped to further Brunei and United Kingdom defence relations. It has also proved many of the operational concepts in anticipation of the arrival of the new amphibious shipping. For Commando Aviation, the new LPH will again see them spending much of their time embarked on HMS *Ocean* supporting the Royal Marines in their amphibious operations, and in the three main operational environments: Arctic, desert and jungle.

US NAVY SEAL SUPPORT

The US Navy SEALs (Sea, Air, Land specialists) number around 2,200 men and are an integral part of the United States Special Operations Command and Navy Special Warfare Command. Commissioned as a unit in January 1962 from the US Navy's specialist Combat Swimmers and Navy Underwater Demolition Teams (UDT), first formed during WW2. The SEALs have gained an enviable reputation as one of the most professional Special Forces units and have been deployed operationally throughout the world, undertaking a diverse range of missions in support of US and Joint Force Special Force operations.

SEAL Teams were deployed to Vietnam (1966–1975) operating throughout the region including the Mekong Delta where their specialist skills were much in demand. SEALs were the first troops ashore during the October 1983, US intervention in Grenada when SEAL reconnaissance teams were deployed ahead of the main force to find and secure landing sites and provide a rescue force to find and protect the British Governor, which they succeeded in doing whilst continually under heavy fire. SEALs also took part in the US intervention in Panama, the Gulf War, Bosnia and many other less publicised actions throughout the world. Like the Royal Marine Special Boat Squadron (SBS), who work closely with their US Navy counterparts, the SEALs are specialists in all aspects of maritime Advance/Special Force operations from deploying covert reconnaissance teams into enemy territory to undertaking clandestine underwater demolition tasks.

SEALs are all trained to conduct Special Forces missions on land and from the sea and all are Special Forces (Airborne) Parachute trained (HALO/ HAHO) and all are capable of conducting land missions, the same as their Army Special Forces colleages. As Navy SEALs, they are also required to be combat swimmers, underwater demolition experts and each undertakes specialist training to become an instructor/ expert in a military Special Forces qualification such as military paratrooping, Seal Delivery Vehicle (SDV) operator, or Underwater Breathing Apparatus (UBA) technician (LARS-5) or a weapons and foreign language expert. SEALs quite literally are experts in Special Forces operations on Sea, Air and Land.

SEALs primary missions are to undertake unconventional warfare, foreign internal defence, direct action, special reconnaissance and counter-terrorism. SEALs are organised into three main components :- SEAL Teams, SEAL Delivery Vehicle (SDV) Teams and Special Boat Squadrons/Special Boat Units (SBS/SBU).

SEAL Teams consist of ten, 16-man platoons, which can operate autonomously and conduct reconnaissance, direct action, unconventional warfare, foreign internal defence and other operations in maritime or riverine environments. This includes advance force, covert reconnaissance and surveillance, demolition of enemy shipping, harbour facilities, bridges and rail-links etc.

SEAL Delivery Vehicle Teams (SVDT) operate and maintain

The US Navy Sikorsky HH-60H Strike Hawk operated by HCS-4/HCS-5 provides CSAR and SEAL support missions. (Sikorsky)

A member of SEAL Team 8 provides cover for his squad from an HH-60H during Helicopter Visit, Board, Search & Seizure (HVBSS) exercises. (US Navy)

various submersible systems that deliver and recover SEALs into hostile areas to conduct reconnaissance and direct action missions. Both SEAL Teams and SDV elements are capable of performing limited shallow water mine clearing operations and are transported by a mother submarine.

Special Boat Squadron/Units (SBS/ SBU) operate a number of coastal and river patrol craft and specialised SEAL insertion craft including the Mark Five Special Operations Craft.

All undertake a multitude of special operations missions, including providing Rescue Teams for Combat Search & Rescue (CSAR) missions supporting Carrier based fast jet operations and rely on helicopters to help insert, extract, re-supply and generally support their many covert/advance force missions.

The USN SEALs can call upon any suitable helicopter to support their missions and often use the large US Marine Corps Sikorsky MH-53Es, USAF Special Operations Sikorsky MH-53J Pave Low 111 and US Army SOF MH-47 Chinooks to provide the range and lift capability to undertake some of their longer-range missions. They also have available the US Navy's fleet of Sikorsky SH-60 Sea Hawks and there are two USN Reserve Squadrons equipped with HH-60H Strike Hawks dedicated to both maritime Combat Search & Rescue (CSAR) and SEAL Special Warfare Support (SWS). SEALs will also soon be supported by the US Navy's version of the Bell/Boeing V-22 Osprey designated HV-22B dedicated to combat SAR, special warfare and logistics support.

A specially modified version of the US Navy Sea Hawk designated the Sikorsky HH-60H Strike Hawk entered service in 1991 to provide support to the SEALs and to undertake CSAR missions and are operated by Helicopter Combat Support Special (HCS) Squadrons HCS-4 at Norfolk, Virginia and HCS-5 at Point Mugu, California. Both are Naval Reserve units with a mixture of full-time and part-time aviators, all experts in providing clandestine support for the SEALs and CSAR including Maritime special operations and low level NVG operations.

The modified HH-60H Strike Hawk retains all the maritime attributes of the SH-60 Sea Hawk including blade / tail fold and can operate from all suitable Navy ships and can be easily air transported in US Air Force C-5/C-17 transports. With a range of 237 nautical miles, the HH-60H is equipped to penetrate hostile territory to either pick-up downed aircrew or to infiltrate or extract SEAL teams. The HH-60H can carry eight fully equipped SEALs and has a crew of four:- two pilots and two crewmen/ gunners/combat swimmers.

The HH-60H has been modified for its clandestine missions and is equipped with a 600lb hydraulic rescue hoist and full aircraft survivability equipment. This includes armoured crew seats, AN/APR-39A(XE) 2 radar and AN/AVR-2 laser Warner's, AN/AAR-47 missile approach Warner's (MAWS), ALQ-144 infra-red jammer, engine exhaust infra-red suppressers (Hover Infra-Red Suppression System/HIRSS), AN/ALE-47 pro-grammable chaff / flare dispensers and night vision goggle-compatible cockpit lighting and is the only US Navy helicopter equipped for NVG operations. The HH-60H has also been

ABOVE: *The Sikorsky HH-60H Strike Hawk is used for maritime CSAR and SEAL support missions and is the only US Navy helicopter equipped for NVG missions.* (Sikorsky)

BELOW: *Navy SEALs conducting Special Warfare insertions using an inflatable boat. These can be thrown out of a helicopter (heli-casting) or, in the case of a Chinook, deployed and recovered from the rear ramp.* (US Navy)

ABOVE: *Navy SEALs exit a USAF MH-53 Pave Low during a CSAR training exercise in Bosnia* (US Navy)

equipped with a digital data-bus and the integration of a pilotage Thermal Imaging/ FLIR system is progressing. For their specialist missions the HH-60H is equipped with the Cubic AN/ARS-6 Personal Locator System (PLS) which is a covert locator system which can home into programmable PLS systems (PRC-112s) carried by fast-jet aviators and Special Forces personnel and gives the helicopter a range and bearing and is coded to individual PRC-112s prior to missions. The HH-60H can also be armed with two M60D machine-guns or two 0.50inch GECAL miniguns.

Like other special operations helicopters, the HH-60H can deliver their SEALs in a number of ways from winching, hover-jumping, fast-roping or heli-casting and regularly practise these methods as well as undertaking long-range insertion/ extractions of SEAL units from the sea or land. Navy HH-60H crews undertake regular CSAR training at the Navy Strike Warfare Center at NAS Fallon in Nevada undertaking realistic mission scenarios including NVG and mountain flying training. The aviators of HCS-4 and HCS-5 undertake some of the most demanding missions flown by today's military aviators (CSAR/Special Forces/Maritime).

The HCS-4/5, HH-60Hs are regularly deployed to sea with Carrier Task Groups providing CSAR cover as well as providing dedicated maritime Special Forces support operating from the sea or land. During the Gulf War HCS-4 and HCS-5 equipped with their new HH-60Hs were deployed to the western deserts of Saudi Arabia based at Al Jouf where they undertook their CSAR missions rescuing downed coalition pilots. During the war phase the two squadrons flew over 461 sorties, many at night on NVGs.

RIGHT: *A Navy SEAL Team member seen dropping from a C-130 Hercules.* (US Navy)

GULF WAR
SPECIAL FORCES SUPPORT

Many of the covert Special Forces missions undertaken during the Gulf War have been made public knowledge by numerous books published by ex-members of the SAS. Many of these missions were dependent on aviation support which was provided by USAF Special Operations Aviation and the RAF's Special Forces Flights (Chinook and C-130 Hercules) which gave these units their mobility and reach and allowed them to conduct their covert activities far behind enemy lines.

During the early hours of 2 August 1990 Iraq began its invasion of Kuwait and seized foreign nationals including British, French and American citizens along with the passengers aboard a British Airways Boeing 747/BA Flight 149 which had been diverted to Kuwait City Airport and was subsequently destroyed. These foreign nationals, including the BA passengers, were taken by the Iraqis and used as 'human shields' at various military and strategic locations throughout Iraq. The invasion of Kuwait resulted in United Nations Security Council Resolution 662 which declared Iraq's annexation of Kuwait to be illegal. This in turn resulted in the formation of the world's largest coalition force comprising over thirty nations which gathered in Saudi Arabia ready to expel Iraq from Kuwaiti territory.

As soon as it was decided to mount a military action to expel the Iraqis, elements of the British Special Forces began arriving in the region. They initially established a training base at Al Minhad, near Dubai in the United Arab Emirates, where they remained until the start of the air war, then moving to their jump-off locations on the Saudi Arabia–Iraq border before the start of Operation Desert Storm. As the numbers of Special Forces began to build up in the region six RAF Special Forces Chinooks were airlifted by USAF C-5 Galaxy aircraft from RAF Mildenhall, Suffolk in October 1990 to undertake intensive desert flying training in preparation for their support role.

The first priority for the Chinook crews was to perfect their low-level desert night-flying capabilities using Night Vision Goggles (NVGs). They needed to be able to fly at heights below fifty feet, even on the darkest of nights, over a difficult undulating desert landscape with little depth perception and the ever-present threat of flying into rising ground. They also needed to perfect their navigation techniques using their newly-acquired satellite Global Positioning Systems (GPS) computers and self-defence suites which included a new Missile Approach Warner and Sky Guardian Radar Warner, along with a new American satellite communications system. Like all the other British helicopters operating within the region, the Chinooks were painted in an all-over desert pink camouflage known as Alkali Removable Temporary Finish (ARTF) paint which when mixed with a detergent, was easily removed. After several long-range NVG training missions the

Chinook flight soon realised that this new desert paint scheme had poor infra-red qualities and almost glowed under NVGs. The Flight Commander, who as a small boy had keenly collected World War Two military models including German night fighters, had the brilliant idea to dab black paint over the top of the desert pink using a similar pattern to the scheme adopted by the German squadrons. The result proved highly successful, blending the Chinook into the desert background both by day and by night. Other notable additions to the Chinooks were the fitting of 7.62mm M134 miniguns and internal Robertson Extended Range Tanks (ERTs).

The first task given to the Special Operations forces while at their training camp in the United Arab Emirates was to research and plan methods of rescuing some of the British and foreign nationals being held in Iraq as human shields prior to the main offensive. A planning team both in London and the UAE began to work on Operation Trebor, receiving daily updates on the location of the hostages. Special Forces Chinooks, C-130 Hercules plus a flight of RN Commando Sea Kings began rehearsing various scenarios. This included a long-range heliborne and para-insertion deep into Iraq with C-130 Hercules landing at a desert strip to set up a FOB inside Iraq with the C-130s acting as a Forward Area Refuelling Point (FARP) to refuel the helicopters. Several C-130 Hercules were fitted with internal fuel tanks, hoses and pumps to act as a mobile petrol station for helicopters or vehicles. This helped to increase the operating range of the helicopters when not fitted with an air-to-air refuelling capability, and was similar to the doomed 'Desert One' FOB during the US military Operation Eagle Claw, the failed 1980 hostage rescue in Iran. The main problem for the planning team for Operation Trebor was that Saddam Hussein, knowing the probability of a rescue mission being mounted, split up the hostages which he then continually moved around. This in turn resulted in the rescue force never quite knowing the number of hostages they could rescue in a single mission. The mission was eventually abandoned, although its name was used again when the British ambassador to the Kuwaiti Embassy was returned.

As the Coalition forces began to formulate their battle plans for the retaking of Kuwait, the role for Special Forces was still unclear. General Norman Schwarzkopf was initially not keen on the use of covert operations behind enemy lines, as many of their intelligence-gathering missions could be undertaken by satellites, Allied aircraft and armoured units once they pushed across into Iraq. The classic role of Special Forces is as an advance force working behind enemy lines to undertake the collation of accurate intelligence information and provide a strategic picture of enemy strengths and dispositions. They are also aptly suited to sabotage, search and destroy missions and

846 NAS Sea Kings refuel on the RFA Fort Grange, returning from a desert training mission in UAE. The Union flags painted on the side of the helicopters had been removed prior to the mission. (Patrick Allen)

ABOVE: *RAF SF Chinooks had their pink desert paint scheme subdued by black paint for their night missions, although at least one SF Chinook retained its original desert colour scheme.* (Patrick Allen)

undertake a secret mission to destroy communications systems around Baghdad.

On the night of 20 January several four-man reconnaissance / observation teams were flown by Chinook into Iraq to keep watch on Iraqi main supply routes. The British area of operations in the Western Desert provided a mixture of terrain ranging from desert sand to lava plateau with rocky outcrops and rock-strewn, uneven surfaces, all difficult to move across on foot or by vehicle. To add to their problems, the weather was proving a challenge with huge contrasts in temperatures and weather conditions. This included below freezing conditions with heavy rain, snow and low cloud and fog, and it was left to individual patrols to decide what kit or vehicles they would take with them.

Two days after the first patrols were deployed by Chinook into the Western Desert, the Iraqis launched another Scud attack on Tel Aviv using their mobile launchers. After consultation with the Americans it was decided that both British and American Special Forces should turn their

BELOW: *One of six RAF Special Forces Chinooks seen at Odiham after the war.* (Patrick Allen)

the bringing about of general mayhem and confusion.

By early January it was finally agreed, after much persuasion, that the SAS/SBS would be given the opportunity to perform two of their classic roles – covert intelligence gathering and sabotage behind enemy lines. They would undertake missions in the Western Desert of Iraq prior to the main invasion, then reconnoitre the Iraqi main supply routes from Baghdad to Jordan. They would destroy any enemy communications systems and conduct hit and run raids against supply lines, create diversions and draw Iraqi troops into the Western Desert region and away from the actual front line. These were classic SF missions and would comprise a number of four- and eight-man reconnaissance teams plus their vehicles, although at a later date larger fighting columns comprising around twenty to thirty troopers equipped with up to ten vehicles including Model 110 Landrovers, Light Strike Vehicles (LSV), Unimog support vehicles and 250cc motorbikes were also deployed.

All these operations required air support either for insertion, resupply or extraction.

On the night of 17 January 1991, to coincide with the air offensive, British Special Forces were flown by RAF C-130 Hercules to a forward holding base located nine hundred kilometres north-west of Riyadh to prepare for their missions. There are now reports of a Special Forces helicopter mission, including RAF Chinooks, flying into the centre of Baghdad and landing in a football stadium on the night of 17/18 January and that the destruction of the early warning radar systems and air raids on Baghdad city that night being a deliberate diversion to hide the helicopter insertion. The helicopters are reported to have been sitting on the ground protected by a large force of Special Forces troops, whilst others went off to

ABOVE: *An RAF Chinook picking up a pair of LSVs during a training exercise. This Chinook was one of six Special Forces helicopters seen without the desert camouflage. (Patrick Allen)*

BELOW: *An RAF Special Forces Chinook creating a sand storm. The paint scheme proved highly effective for their clandestine missions. (Patrick Allen)*

attention to 'Scud hunting' and destroy these mobile launchers. On 23 January two RAF Chinooks, together with a team from the SBS, executed one of the most successful missions of the war. Having gained information from the French Secret Service on the location of the main Iraqi military communications network system and cable route running between Baghdad and Basra and then to their front lines, this small raiding party flew to within sixty miles of Baghdad and destroyed the system. The two Chinooks landed alongside the main Baghdad to Basra road where the communications lines had been sunk underground and dropped off the SBS team who then proceeded to dig down to the cables. The two Chinooks in the meantime moved away to a nearby clearing and, protected by a small team, waited on the ground with their rotors disengaged but with their auxiliary power units running to help reduce engine noise. The SBS party dug down, exposed the communications cables, cut out a section to take away for analysis and then set a demolition charge. The whole mission was undertaken without any drama and was a complete success, and the team returned with a Basra signpost to present to the Commander-in-Chief.

By 29 January 1991 the American Special Forces were keen to enter the Scud-hunting role and it was agreed that the British would patrol the territory south of the main supply route running between Baghdad and Amman, the Americans patrolling to the north. The first American Special Forces deployment into the area they called Scud Boulevard took place on 8 February when teams were inserted by USAF Special Operations MH-53Js. Fitted with a FLIR and terrain-following radar the MH-53s could operate in all weathers and had already flown a number of missions for the British Special Forces plus several Combat SAR missions to recover downed aircrew from deep inside Iraq.

On 26 February a Special Forces planning meeting took place to discuss the completion of Operation Trebor and missions within Kuwait City. The SBS were flown into Kuwait City on 27 February, and after an initial delay Operation Trebor

RAF Special Forces Chinooks operated almost exclusively at night with crews wearing NVGs supporting SAS/SBS missions. (Patrick Allen)

The RAF purchased the M134 Minigun for their Special Forces Chinooks during the Gulf War. It was used in anger on at least one occasion. (Patrick Allen)

began the next day when three Commando Sea Kings picked up their SBS troops from Kuwait City Airport, where they had been flown in by RAF C-130 Hercules. These three Sea Kings were later joined by three RAF Special Forces Chinooks which had been trying to get up to Kuwait from Saudi Arabia but were having difficulties flying around in the black smoke caused by the oil-well fires. The mission involved two Sea Kings dropping observation teams onto the roof of two nearby high-rise buildings which overlooked the embassy, with a third Sea King and a pair of RAF Chinooks fast-roping the teams onto the Embassy roof: the third Chinook, having collected the British ambassador, flew to a nearby safe landing area. There were no losses or damage to any of the RAF aircraft during their covert missions into Iraq supporting Special Forces. The only blemish on the RAF's record book occurred after hostilities had finished and involved a Special Forces Chinook which landed at an Iraqi military airfield. With its rotor downwash, this disturbed an unexploded JP233 bomblet, causing damage to the rear section and rear wheels.

HUMANITARIAN OPERATIONS
KURDISTAN

Over recent years there have been a number of operations in support of United Nations peacekeeping, protection and humanitarian assistance missions; the two most recent involved multinational missions in Somalia and Bosnia. It was at the end of the Gulf War in April 1991 that the first major UN humanitarian mission took place on which many of today's operations and procedures are based.

Immediately after the the Gulf War a massive Allied effort was launched to provide humanitarian assistance to thousands of Kurdish refugees who were under direct threat from Saddam Hussein's forces and who had fled into the mountainous region between Iraq and Turkey. The region was inhospitable and difficult to reach, and the only way to deliver urgently needed supplies was by air. This was initially carried

USMC CH-53 seen landing alongside a CH-46 to be loaded with UNHCR supplies at Silopi, the main Forward Operating Base for the Allied helicopter force.
(Patrick Allen)

US Army AH-64A Apaches refuel at the FARP at Silopi. They provided escort and protection for the Allied helicopters undertaking UNHCR missions in Iraq. (Patrick Allen)

out by stores drops from C-130 Hercules aircraft as preparations were made for the deployment of a huge Allied helicopter force along with ground troops who would first feed the refugees and help prevent further deaths, then establish a safe haven within northern Iraq for the returning Kurds. Among the first helicopters to arrive in eastern Turkey were nine RAF Chinooks, half of which had self-deployed from their home base at RAF Odiham, Hampshire with the advance party arriving at Diyarbakir on 9 April 1991. The first Chinooks arrived on 14 April. Taken from Nos 240 OCU (27 Sqn), 7 and 18 Squadrons a number of these Chinooks had been diverted from their return to the UK from Saudi Arabia

aboard the MV *Baltic Eagle*. Already operating within the region and based at Batman near Diyarbakir were USAF Special Operations MH-53J Pave-Lows from the 21st SOS, plus a number of HC-130 and MC-130E Combat Talons which had been undertaking Special Operations and Combat SAR missions out of Batman during the Gulf War, and which were now well placed and well equipped to undertake humanitarian missions in support of the Kurds.

Arriving at about the same time as the RAF Chinooks were three US Navy CH-53Es from VC-4 based at Sigonella, Italy. They were followed by the helicopters from the USMC 24th Marine Expeditionary Unit operating twelve CH-46E Sea

Knights, four CH-53Es, three UH-1Ns and four AH-1J/Ws. US Army helicopters were next to arrive during the second week of April, comprising twenty CH-47D Chinooks from E and D Companies of the 502nd AVN, plus thirty-six UH-60 Black Hawks from 11th Armoured Cavalry, the 159th Ambulance Company and H Company, 4th Brigade, 8th Infantry plus AH-64A Apaches from 6/6th Cavalry. These helicopters were joined by eleven Sea King HC4s from 846 NAS and Lynx and Gazelles from 3 Commando Brigade Air Squadron, along with Dutch Air Force Alouettes, French Super Pumas, Italian and Spanish Chinooks, Turkish Hueys and Black Hawks and German Army UH-1D Hueys and CH-53Ds.

As one of the first helicopter units to arrive in theatre the RAF Chinooks began flying missions on 14 April operating from Diyarbakir and forward deployed to Silopi on the Turkey–Iraq border, where within a few weeks the entire Allied helicopter force would be located. Using their 800-gallon internal extended-range tanks the RAF Chinooks were able to operate in the more remote mountain regions in the east of Iraq as far as Hakkari and Yuksekova, close to the Iranian border, delivering urgently needed food and medical supplies. In those early days the logistics chain had not yet been established and the Chinooks were operating in these remote

LEFT: *USN CH-53E seen with a US Army Black Hawk, only part of the huge helicopter force assisting in the humanitarian missions in Northern Iraq.* (Patrick Allen)

BELOW: *USAF Special Operations MH-53J Pave Lows picking up troops for the push into Northern Iraq to create the Kurdish 'Safe Haven.'* (Patrick Allen)

USAF MH-53J Pave Lows were one of the few helicopters capable of operating during poor weather in Northern Iraq. The MH-53s operated from Batman undertaking SF/CSAR missions during the War. (Patrick Allen)

areas on their own, distributing food (mainly bread), clothing and medical supplies donated by local Turks until more UNHCR relief supplies became available.

Landing at the mountain refugee camps proved extremely hazardous in those early days. As refugees rushed towards arriving helicopters they seriously risked decapitation on the front rotor disc of the Chinooks or being crushed or injured. Pilots were forced to hover above the crowds with the crews throwing bread and supplies off the rear ramp until safe food distribution areas could be organised. As more and more assets became available US troops from the 10th Special Forces Group along with Royal Marines of the Mountain and Arctic Warfare Cadre (M&AW) were deployed into the mountains to give much needed medical assistance and to establish and run helicopter landing sites and local food distribution. The only food readily available and easily transported at this time was in the form of US military Meals Ready to Eat (MREs), which many Kurds refused to eat. Once the UNHCR supply chain and distribution network had been established, a more natural diet of flour, lentils, sugar, rice and salt became available and could be lifted in external loads by the helicopters.

As the logistics chain improved and UNHCR aid started to arrive in bulk, Chinooks and CH-53Es began to maximise their enormous lifting capabilities. With larger quantities of supplies arriving at Silopi, British Joint Air Support Unit (JHSU) and Mobile Air Operations Teams (MAOTs) began operating a load park at Silopi where they organised the unloading of lorries laden with supplies then repackaged them for helicopter transport, mainly in cargo nets. These netted loads

allowed Chinooks to maximise on every mission, carrying up to three nets per mission to three separate landing sites. In a single day RAF Chinooks were moving over 130 tons of cargo and undertaking 150 km transits from Silopi into the high mountains. Between 14 April and 4 May the nine RAF Chinooks totalled over 380 sorties, completed 637 flight hours lifting over 1,300 tons of relief aid, and undertook 179 medevacs. They also moved over 1,400 troops, carried 496 refugees and shifted 110 tons of marine freight.

Most of the refugees had camped in the mountains, which rose to around 8,000 ft. The nearest military airfield was Diyarbakir about 175 miles to the west, and which became the major airhead (hub) for both incoming supplies and the main maintenance base for many of the helicopters. Silopi, situated on the Turkey–Iraq border close to the mountains, was ideally situated for the Allied helicopter force and a large field outside the town was purchased. This allowed Incirlik and Diyarbakir to remain as the two major airheads, with the helicopters and their crews forward deployed to Silopi. In late April the field at Silopi saw more helicopter movements in a single day than London Heathrow Airport, with the British refuelling point handling over seventy refuels involving over 60,000 gallons of AVTUR. The American refuelling point situated alongside also broke all records.

The Chinooks and USMC CH-53Es were among the few helicopters powerful enough to carry their heavy netted loads directly over the tops of the mountains in the hot and high conditions of the region. Many of the refugee camps were in excess of 8,000 ft and pilots needed to be aware of the hazards associated with hot and high density altitude mountain flying, particularly when flying with heavy internal loads. This required safe single-engine performance at all times. With an external load, if an engine fails the pilot can 'pickle off' the load and instantly reduce weight to regain single-engine performance. With the region noted for large temperature changes in short periods, there was also the risk of extreme meteorological conditions such as mountain wave turbulence, dangerous downdrafts and reduction in helicopter performance due to decreased air density, which is caused by both altitude and temperatures (80°F at 8,000 ft). This is particularly critical during hovering or landing when engines and rotors are working at their hardest. At least one USN CH-53E crashed high in the mountains when it lost tail rotor authority during a landing.

While the humanitarian helicopter operations were continuing, Operation Haven was getting underway with the arrival of Allied troops including Royal Marines, US Army, USMC and French forces. The Allies needed to get the Kurds off the mountains and back to their homes in the north of Iraq as quickly as possible. It was decided to establish a safe haven of approximately seventy to a hundred kilometres along the Silopi valley bowl in northern Iraq, which would allow the Kurdish refugees to leave their mountain camps. Initially protected by Allied troops, the responsibility for their safety would be handed over to the United Nations once the safe haven had been fully established. With typhoid and cholera outbreaks increasing in the mountain camps and with the onset of summer, time was extremely short. The Kurds had been melting snow as their water source. With the snow rapidly disappearing, the helicopters could not keep up with the increased demand for water.

During both Operation Provide Comfort and Operation Haven Allied helicopters had the protection of fighter top cover and close air support, and they were in continual contact with AWACS aircraft which provided command and control and, more importantly for the helicopters operating in this mountainous region, a Flight Following service. At night this 'eye-in-the-sky' was provided by USN Hawkeyes. If a helicopter got into trouble a quick radio call to AWACS or Hawkeye would be enough to bring down the patrolling F-15s, F-16s and A-10s to provide assistance. If a helicopter was

US 10th Special Forces troops seen working with an 846 NAS Sea King at a mountain landing site, moving Kurdish families off the mountains. (Patrick Allen)

An RAF Special Forces painted Chinook shows the capabilities of the Chinook, lifting a triple netted load of UNHC supplies which were flown high into the mountains to the Kurdish refugee camps. (Patrick Allen)

shot down or crashed a Combat SAR mission would be immediately launched by the USAF Special Operations MH-53J Pave Lows with A-10s and AWACS providing a watchful eye until the crew or troops were safely recovered.

In the event the operation was unopposed and a complete success. Within a few days the safe haven had been established and helicopters were busy ferrying the refugees off the mountains in large numbers. By the end of June the majority of the original Allied military force had completed their mission and left the region, and its management was turned over to the United Nations. The Allies continued to provide an air presence in the region, flying regular combat air patrols to police the northern Iraq no-fly zone under the new name 'Operation Warden'.

COMBAT SAR

Combat Search and Rescue is a growth industry for today's military planners and the role, once the province of search and rescue units, is now mainly undertaken by Special Operations Aviation. Special Forces and CSAR missions are similar in that they both require covert operations within a high-threat environment, usually behind enemy lines to insert, resupply or extract Special Forces, or to rescue downed aircrew or hostages. Captured military aircrew or Special Forces personnel have never been so highly prized for their military and political value. During the Gulf War captured Coalition aircrew were displayed by the Iraqis across the world's media in an attempt to influence public opinion at home and abroad. In Bosnia, the unknown fate of two French pilots downed by the Bosnian Serbs near Pale was used as a negotiating lever.

Air operations during the Gulf War and in Bosnia saw air power and precision bombing used to inflict military and political pressure upon the enemy. Used as a means to postpone the deployment of ground troops, these air operations involved co-ordinated bombing campaigns to destroy specific military targets, communications, transport and basic infrastructure. Use of air power proved a success in the Gulf War and was repeated in Bosnia in 1995 as NATO aircraft undertook a bombing campaign aimed at Bosnian Serb roads and bridges, communications and military installations. This helped to force the Bosnian Serbs back to the negotiating table and resulted in the ceasing of hostilities and the eventual deployment of the NATO-led IFOR/SFOR. Having

been shown the 'hard stick' there was always the constant threat of further air strikes by NATO aircraft if the Bosnian Serbs did not comply.

The new emphasis on the use of air power as the first military action has centred attention on the safety of the aircrews involved, as they often have to overfly unfriendly territory. In turn, this has produced a renewed enthusiasm for effective CSAR in terms of the recovery of downed aircrew from within a hostile or enemy-held territory. The death, capture or public parading of captured aircrew remains unacceptable to the military and to the general public in the West. Their safe return, preferably without having been captured, remains a priority. CSAR operations have always been difficult and dangerous, requiring specially trained aircrew and dedicated aircraft and helicopters. The CSAR mission is so specialised that many of its operational procedures are identical to those used by Special Forces, i.e. the long-range, covert infiltration or exfiltration of advance forces.

The similarity between CSAR and Special Forces operations resulted in the US military amalgamating its CSAR assets within the US Special Operations Command, which was activated on 16 April 1987. USAF CSAR units, formally known as the Aerospace Rescue & Recovery Service (ARRS), are now part of the USAF Special Operations Command and have been retitled Special Operations Squadrons (SOS), working alongside US Army and US Navy Special Operations units. The USAF have also equipped Reserve and Air National Guard Air Rescue Squadrons which operate the Sikorsky HH-60G Pave Hawk,

Helicopters and C-130 Hercules are an ideal combination for CSAR. (Lockheed)

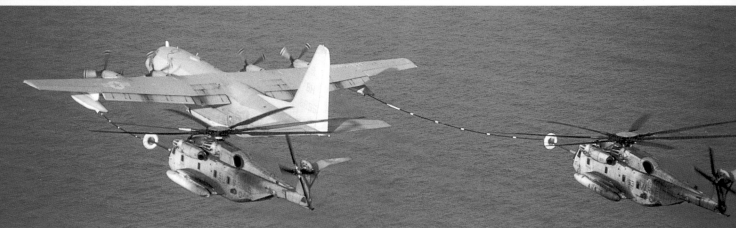

TOP: *A C-17 Globemaster 111 seen landing at a desert strip. It could be used as a FARP for future CV-22B Osprey CSAR/SOF missions.* (Boeing)

ABOVE: *A pair of CH-53Es from HMH-464 refuel from a USMC C-130 during a TRAP/CSAR training mission.* (Patrick Allen)

and the US Naval Reserve operating the Sikorsky HH-60H Hawk, both dedicated to CSAR.

Operating over unfriendly territory, with the ever-present threat from modern airborne surveillance and sophisticated enemy air defence systems, CSAR missions are some of the most demanding and dangerous. New long-range stealth aircraft like the Lockheed F-117 Night Hawk and Northrop Grumman B2A Spirit, designed to operate deep within enemy-held territory with little or no support, has made the combat range of modern aircraft almost unlimited. This range needs to be matched by CSAR assets which have the added require-ment that they usually need to operate at low level and must hover or land at the halfway point to recover personnel.

Operating over enemy territory remains one of the most challenging missions a helicopter pilot can undertake. Unlike most Special Forces missions when the time and place can normally be chosen, CSAR missions are initiated at the time

and place of the shoot-down. To make matters worse, the enemy is usually aware that an aircraft is down and will also be looking for the crew. They also know that unless the aircrew is captured, a CSAR mission will almost certainly be mounted. The American military invented and perfected CSAR procedures during the Vietnam War and today, other NATO nations are beginning to expand their CSAR capabilities including the UK, France and Italy. There are several types of CSAR missions. The Americans with their military power are capable of launching a rescue mission which will deliberately engage an enemy force to recover downed aircrew. The methods favoured by NATO countries and by the US Marine Corps is a CSAR procedure known as the Tactical Recovery of Aircraft and Personnel known as TRAP. This uses all the same basic procedures as CSAR, but the rescue force will try and rescue their survivor without engaging enemy forces. This method makes more use of Escape and Evade procedures and the covert insertion of rescue teams/Special Forces to collect downed aircrew. Like other CSAR missions the rescue force of helicopters and Special Forces is protected by a Combined Air Operations (COMAO) package which includes fixed-wing Rescue Combat Air Patrol (RESCAP) and rescue escort (RESCORT), suppression of enemy air defence (SEAD) aircraft, tankers and other escort and command and control assets. To help counter some of the disadvantages of being in hostile territory, aircrew and Special Forces personnel receive considerable training which includes combat survival and escape and evasion (E&E) training and CSAR code procedures. This training is designed to get aircrew away from their crash site and allow time for a CSAR mission to be mounted.

Detailed planning and co-ordination is a requisite for modern CSAR missions, taking full advantage of modern technology including detailed reconnaissance (satellite mapping) AWACS and, if time permits, a detailed rehearsal of the mission in flight and mission simulators.

Special Operations Command

The failure of 'Operation Eagle Claw', the mission to recover fifty-three US hostages from Tehran on 24 April 1980, and the disaster at 'Desert One', the covert Forward Operating and Refuelling Base (FARP) established in the desert 265 miles south-east of Tehran which resulted in the death of eight people and the cancellation of the mission, forced the US

The USMC AH-1W Super Cobras provide close-in escort for USMC CH-53Es during TRAP/CSAR missions. They were used in Bosnia during the Basher mission.
(Patrick Allen)

military to form a dedicated Special Operations Command.

There were many reasons for the failure of Operation Eagle Claw. These included unexpected weather conditions, poor communications, non-specialist helicopters and aircraft from different units and services not having worked together as a combined unit. The Navy RH-53D helicopters had been quickly re-roled as Special Operations aircraft and were flown by US Marine Corps pilots who were unfamiliar with this particular model of CH-53. In addition they were equipped with new and unfamiliar navigation equipment. The troops for the mission, including the security party at the Desert One refuelling site, were from Delta Force and the Green Berets plus other US Army Special Forces units which, again, had not worked together on this type of scenario. Of the eight RH-53Ds departing USS *Nimitz*, their numbers reduced to five due to mechanical failures on the transit to Desert One. Although the USN RH-53Ds had an aerial refuelling capability, the Hercules C-130s used in the mission had no such capability. This limited their range and necessitated the setting up of a FARP at Desert One before the launch of the mission into Tehran.

The RH-53Ds certainly had a difficult transit to Desert One. Once overland they almost immediately went onto instrument flying, having to climb high above sandstorms which required the pilots to fly continually on their instruments and use 'dead reckoning' before dropping down to Desert One. With the loss of the three RH-53Ds, the decision was made to abort the mission. The five remaining RH-53Ds and four C-130s waiting on the ground with their crews wearing full-face ANVIS 5 NVGs had to undertake an unplanned and unrehearsed reposition on the FARP so that the helicopters could

A Heavily armed USMC CH-53E seen in Turkey during Operation Safe Haven. These aircraft were ready to undertake TRAP/CSAR missions. (Patrick Allen)

refuel from the C-130s prior to their transit back to the USS *Nimitz*. This unplanned repositioning resulted in an RH-53D clipping the tail of a C-130 and causing a catastrophic fire.

The failure of the mission resulted in the setting up of the Holloway Commission to review the events of Operation Eagle Claw and recommend ways to prevent this ever happening again. The commission's conclusions resulted in the amalgamation of the USAF ARRS units with Special Operations Aviation units which would regularly work with Army, Navy and other Air Force Special Operations units establishing common Standard Operational Procedures (SOPs) to perfect and expand their Special Forces and CSAR capabilities. It was also recommended that both the Army and Air Force be equipped with dedicated Special Operations aircraft and helicopters designed specifically for night, long-range, low-level, covert, all-weather operations. The US Army began establishing their own Special Operations Aviation unit known originally as Task Force 160. They re-formed to become the 160th Special Operations Aviation Regiment (Airborne) and were initially equipped with modified MH-47D Chinooks, MH-60 Black Hawks and MH/AH-6 'Little Birds' prior to the arrival of their dedicated fleet of MH-47E Chinooks and MH-60K Black Hawks. The USAF, under Operation Frequent Wind, began to upgrade and modernise their fleet of HH-53C Pave Knife CSAR helicopters under Project Pave Low to become Special Operations MH-53J Pave Low 111 helicopters. There were also modifications to their C-130 Hercules, MC-130 Combat Talons, HC-130 Combat Shadow and AC-130 Spectre gunships, plus additional MH-60G Pave Hawks and HH-60G Pave Hawk CSAR helicopters.

By the time of the Gulf War CSAR was under the jurisdiction of Special Operations Command with the USAF operating dedicated Sikorsky Special Operations Forces (SOF) MH-53J Pave Low 111s and MH-60G Pave Hawks along with their HC-130 Combat Shadow tankers, MC-130 Combat Talons and AC-130 Spectre gunships. The US Navy now relied on their own HH-60H Rescue Hawks operated by their Special Helicopter Support Squadrons. The USMC relied on their own Special Operations-capable Tactical Recovery of Aircraft and Personnel (TRAP) teams. During the Gulf War USAF Special Operations MH-53J Pave Lows MC-130 Combat Talons and HC-130 Combat Shadows operated from Saudi Arabia and eastern Turkey near Batman undertaking CSAR and Special Forces operations. It has been reported that their MH-53Js were used for over sixty Special Forces operations inside Iraq, working for both the US and British Special Forces, including one MH-53J picking up an Iraqi officer taken prisoner by the SAS. Many of these missions were in support of the 'Scud busting' operations undertaken by British and American Special Forces.

USAF SOF MH-53Js carried out the first mission of the war when at 0238 hours on 17 January 1991 a pair of MH-53Js acted as Pathfinders. Using their terrain-following/avoidance radars and thermal-imaging systems they escorted eight AH-64A Apaches of the 101st Airborne (Air Assault) Division across the Saudi Arabia–Iraq border deep into the Western Desert of Iraq to take out two Iraqi early warning radars and communication systems to create a 'black corridor' prior to the first wave of Allied aircraft entering the country at the start of Operation Desert Sabre. The MH-53Js also undertook several CSAR missions inside Iraq. On 21 January 1991 a pair

Fast jets proved RESCORT/RESCAP support for helicopters on CSAR missions. The RAF Tornado GR1, seen here refuelling from an RAF VC10 tanker, could provide a RESCORT capability if needed. (Patrick Allen)

of MH53Js from the 20th SOS escorted by A-10 'Sandys' from 354th TFW flew on a daylight CSAR mission 150 miles into Iraq to pick up a downed US Navy F-14 Tomcat pilot.

By the time of the Gulf War aircrew were being issued with updated PLB/TACBEs including the PRC-90 and improved PRC-112s. Today's high-powered digital Personnel Locator Systems (PLS) TACBEs, such as the PRC-112, allow the user to talk to AWACS or rescue aircraft on a pre-programmed encrypted frequency which uses burst transmissions to reduce the risk of interception by the enemy. Capable of being programmed with their own individual codes, the more advanced transmitters can be automatically interrogated by CSAR/SOF/AWACS aircraft to verify their authenticity. Using burst transmissions the PLS/TACBE can be automatically interrogated by the CSAR/aircraft/helicopter which gives the helicopter a steering cue (bearing and range). This allows the CSAR helicopter/C-130 to remain in terrain masking, only having to pop up quickly to interrogate the PLS/TACBE to gain an update for the aircraft's navigation systems. TACBEs and the PRC-112 are standard equipment on today's Special Forces aircraft and are used by aircrew and Special Forces personnel. Today, PLSs used by pilots and Special Forces have an embedded GPS receiver providing the rescue teams with an effective over-the-horizon operational communication range of hundreds of miles with an automatic advanced encrypted interrogation response protocol allowing accurate one-pass pick-up under the worst conditions. The system also allows two-way encrypted communications between Special Forces, downed pilot and rescue force. These systems, which are pocket-sized, provide outstanding location accuracy and communications security with over 3,000 operational channels, two-way UHF voice radio and an effective range in excess of 300 km. The embedded GPS system can also be used as a navigational aid for downed aircrew and Special Forces should they undertake E&E.

Escape and Evasion

Although CSAR missions can be tailored to suit the operational and threat environment, sometimes the risks can be too

great and require downed aircrew and Special Forces to undertake escape and evade tactics. Today, pilots are well equipped and trained for survival and will have been carefully briefed on what to expect should they land up on the ground with respect to climatic conditions and enemy forces and dispositions etc. Once on the ground they make their way to pre-selected E&E Lying-Up Points (LUPs) where either Special Forces will have been inserted to receive them, or where they must await the arrival of a CSAR helicopter at a pre-determined time. The pick-up will only take place if the CSAR team can authenticate the identity of the downed aircrew. This usually involves a detailed recce of the area and the use of pre-selected code words or signals along with a verbal interrogation using the TACBE to obtain answers to pre-selected questions which are known only to the aircrew and have been lodged with the rescue package. The pre-programmed locator beacon code, which is given to each individual aircrew, must also be verified. At night, when the CSAR crews are using NVGs, the downed pilot can use an IR 'Beta-light' or Cyalume to flash a code if required. Prior to any mission aircrew and Special Forces are given a number of CSAR codes. These will help the Rescue Force authenticate them. Prior to missions aircrew are given a 'Bull's Eye' which is Lat\Long position within the operational area. Downed aircrew always give their position from the 'Bulls Eye'. The Ramrod code is a ten letter word with each letter of the word representing a number ie. BLACKHORSE/ 0123456789. The Ramrod changes daily. Aircrew give their position using the Ramrod code. There is also a final daily code which comprises a day number, letter and word for example: Day 13, No 76, the letter H and the word PAINTBALL along with a duress word such as Jackal which a survivor can put into his radio call to let the rescue force know he is speaking under duress. The survivor will be asked several questions to check authentication. There is also a final CSAR ISO Prep Form. This is the answer to several personal questions which the aircrew will have lodged with his unit prior to operations such as his wife's middle name or a birthmark etc. If photographs are available the Rescue Force will have them.

The USMC are also well practised in the recovery of personnel from behind enemy lines and have designed and adapted their own Tactical Recovery of Aircraft and Personnel (TRAP) missions to suit most operational conditions. The Bosnian conflict brought home the difficulties of recovering downed aircrew while operating over highly populated areas with no discernible front line and the ever-present risk of landing right on top of the enemy or among a hostile population. On 2 June 1995 a USAF F-16 pilot, Captain Scott O'Grady (callsign Basher 52) from the 555th FS/31st FW, was shot down over Banja Luka and remained hidden from the Bosnian Serbs for six days. Once it had been confirmed that O'Grady was both alive and had evaded capture, the USMC initiated a TRAP mission to recover the pilot. TRAP missions can involve the co-ordination of large COMAO packages of fixed-wing aircraft and helicopters from different services (USMC/USN/USAF/US Army). The Marines have tailored these COMAO packages to suit various threat scenarios based around their TRAP Rapid Response Planning Programme, which is itself based on standard operational procedures with the force package dependent on both the type of mission and expected threats. This allows them to launch a tailor-made TRAP mission adapted to the individual operation, within six hours of receipt. TRAP missions are only launched if certain criteria are known: location of the survivor(s), confirmation that they are alive and have not been compromised or are in danger of capture, and full authentication of the survivor(s) identity. Once these basics are known, the TRAP planning team can start to estimate the size of the mission and begin getting all the personnel, aircraft and agencies together.

The Scott O'Grady Basher 52 mission was based around a standard TRAP 'DELTA' operation lasting for two and half hours, allowing a 100nm radius of action with thirty minutes on scene for a ground search. It comprised a standard package of two Sikorsky CH-53Es with a thirty-strong TRAP security team which included Forward Air Controllers (FAC) and medics, two escort AH-1W Cobras, four AV-8B Harriers

(CAS) plus AWACS airborne command and control plus fighter top cover and SEAD/HARM aircraft.

The USMC favour their big CH-53Es over the smaller MH-53Js of the USAF as each CH-53E is capable of recovering the entire complement of the mission package (troops/aircrew) without regard to capacity, fuel or weight restrictions should one of the aircraft or helicopters go down *en route*. This reduces the risk of leaving people behind if a helicopter or aircraft is shot down. The big CH-53Es also have the necesssary speed and endurance to overcome many unforeseen events should the mission become difficult.

Once at the survivor's location the Cobras clear the LZ for the incoming CH-53Es while the AV-8Bs provide cover should a Cobra need to call down an air strike. Once the CH-53Es have landed at the LZ, the TRAP security team secures the LZ while the medics attend to the survivor(s), if required. Should the LZ become 'hot' the Cobras and AV-8Bs plus back-up from additional fighter top cover engage the enemy while the CH-53s pick up their personnel/survivors and depart for an aerial refuelling RV with a tanker, then home.

CSAR and Special Operations have never been easy. On today's high-threat battlefields these types of missions are becoming increasingly demanding, providing the ultimate challenge to both the rescue teams and downed aircrew. Aerial refuelling, digital satellite communications, terrain-following radars, AWACS command and control, Special Operations C-130s and dedicated Special Forces/TRAP security teams together with realistic multi-agency training have helped to reduce the risks. These risks are the same should the mission involve the recovery of one pilot or 100 hostages. Combat rescue of aircrew from enemy territory along with Special Operations will benefit from the arrival of new aircraft such as the Bell/Boeing V-22 Osprey, which is destined for the USAF Special Operations (CV-22), USMC (MV-22) and USN (HV-22), although today's Special Operations aircraft and helicopters will continue to provide an invaluable service for many years ahead.

An RAF Chinook seen operating in Bosnia. They are highly capable of providing TRAP/CSAR support to SFOR troops if required. (Patrick Allen)

AIRBORNE INTERVENTION
AND HOSTAGE RESCUE

The Israelis set the benchmark for the successful use of airborne and air-land techniques when they pulled off their famous 'Operation Jonathan' Entebbe raid on 4 July 1976. This operation showed the world that with the intelligent and imaginative use of air transport assets it was possible to conduct a decisive and effective air-land operation, far from home with little time for planning. Many of the techniques devised and used by the Israelis during this operation have since been taken up by today's airborne brigades when undertaking similar missions.

In today's uncertain strategic environments, combined with reduced defence expenditures, military planners are turning more towards airborne, airmobile and amphibious forces to provide the flexibility, mobility and global reach required. These forces need to be capable of working either independently or as part of a larger combined force. They should be light, highly mobile and capable of worldwide deployment to undertake a wide spectrum of roles ranging from direct intervention to humanitarian relief. The British have three such units: 5 Airborne Brigade, 24 Airmobile Brigade and 3 Commando Brigade, Royal Marines. These three have recently been combined to form the UK's new Joint Rapid Deployment Force (JRDF). Each of the three brigades is capable of working individually, together, or as part of a much larger NATO Rapid Reaction Force.

5 Airborne Brigade

The UK's main airborne intervention force is 5 Airborne Brigade, part of 3 (United Kingdom) Division. The brigade is part of both the new UK JRDF and the UK's contribution to the Allied Command Europe Rapid Reaction Corps (ARRC). The brigade retains a high availability for national contingency operations and is the UK's spearhead force to undertake an intervention role for the UK's Services Protected Evacuations (SPE), and regularly practises this role. No. 5 Airborne Brigade

embraces the unique capabilities of parachute, air-land or airmobile assault using both rotary- and fixed-wing assets to undertake seize and hold operations, to establish an entry point or airhead to a 'theatre of operations', to conduct surgical actions for the evacuation of nationals, and to handle the rapid build-up of heavier follow-on forces. To this end, the brigade includes two battalions of the Parachute Regiment in the parachute role and two airportable infantry battalions (one is always drawn from the Royal Gurkha Rifles). These are designated 'airborne infantry battalions' and undertake air-land operations. The brigade maintains its Leading Parachute Battalion Group (LPBG) at five days' notice to emplane for operations worldwide. Along with these four battalions the brigade maintains a Pathfinder platoon with an ability to deploy ahead of a main force to provide reconnaissance, mark drop zones and helicopter landing sites, and to undertake other missions assigned to advance forces, in a similar manner to the Royal Marine's Brigade Patrol Group (BPG).

The Brigade can launch a parachute operation to conduct an overhead or offset assault by parachute drop, a TALO undertaking an air-land assault on an airhead, or a combined parachute and air-land assault by day or by night. Furthermore, both parachute and air-land assaults can be launched from the home base, a Forward Mounting Base (FMB), or a friendly nation's airfield.

A typical exercise scenario usually involves the brigade going to the assistance of a friendly country which is under direct threat from a neighbour, or has asked for assistance due to a complete breakdown in civil order. This can also include the safe evacuation of non-combatants, British expats and other nationals friendly to the UK. To further complicate the scenario there can be the added difficulty of securing or retaking an embassy building and securing the

US Airborne troops are turning towards Air Assault as their new role which is similar to the role of these US Army 101 Airborne Division troopers, seen setting up a tactical SAT Comm system. (Patrick Allen)

ABOVE: *Night operations are a must for any intervention role. A Commando Sea King is seen through NVGs operating in Bosnia.* (Patrick Allen)

RIGHT: *US Army Black Hawks from 101st Airborne Division.* (Sikorsky)

safety of embassy staff. In the worst-case scenario this can also include the finding and recovery of any hostages or nationals being held against their will.

Exercise Purple Star 96

'Exercise Purple Star 96', which took place on the east coast of the USA in the spring of 1996, was the first major work-up for the UK's new JRDF, and this was followed in 1997 by the Royal Navy's 'Ocean Wave 97' deployment to the Far East. Purple Star 96 was planned to practise the strategic deployment of elements of this force, to exercise command, control and the support of follow-on forces (National Contingency Force) in concurrent and combined operations with US coalition forces under a Combined Joint Task Force (CJTF) command and control concept. The exercise also involved a wide range of combined UK/US airborne, airmobile and amphibious assault misisons.

The UK's 5 Airborne Brigade are fully capable of operating at night and the flexibility, reach and firepower provided by these airborne units have proved ideal for a multitude of roles from hostage recovery to full-scale intervention. The UK's airborne forces rely on RAF transport and support helicopters to provide them with their operational versatility and reach. This will be further increased when the RAF receives its new-generation C-130J Hercules and Future Large Aircraft along with new support helicopters like the Chinook and EH101 Merlin HC3, and the British Army's WAH-64D Longbow Apache.

ABOVE: *An RAF C-130 dropping cargo pallets during an Airborne Intervention mission.* (Patrick Allen)

BELOW: *RAF C-130s flying a Tactical Air Land Operation Mission (TALO). This type of airborne air–land capability will continue to be important.* (Patrick Allen)

TOP: *RAF Chinooks transit to a FOB site during a 5 Airborne Brigade exercise.* (Patrick Allen)

ABOVE: *US Army Airborne Rangers rappelling from a Black Hawk.* (Patrick Allen)

RIGHT: *The Parachute Regiment will continue to retain a parachute capability plus a new Air Assault capability. They are part of the newly created UK 16 Air Assault Brigade. Paratroopers are seen jumping from a Chinook.* (Patrick Allen)

ARCTIC FLYING

Royal Navy Commando Sea King aircrew are well trained to undertake missions in extreme climatic conditions. There are two front-line commando squadrons 845 and 846 Naval Air Squadrons, plus a training squadron, 848 NAS, all based at RNAS Yeovilton in Somerset, UK. These RN commando squadrons provide the helicopter support for 3 Commando Brigade, Royal Marines.

Every winter since 1969, Royal Navy Commando aircrew and maintenance personnel have received specialist training at the RN 'Clockwork' detachment at Bardufoss, a Royal Norwegian Air Force base located 100 miles inside the Arctic circle. The purpose of the exercise is to undertake cold-weather and mountain warfare training which includes Arctic survival, flying and military training. The Royal Navy believes that learning to live and operate effectively in severe cold-weather conditions, and to overcome an inhospitable operational environment, provides invaluable experience for both aircrew and ground personnel. This not only enhances their operational capabilities when supporting 3 Commando Brigade, Royal Marines and other specialist troops, but also proves to be invaluable in enhancing their operational effectiveness wherever in the world the RN Commando helicopter squadrons find themselves.

The operational training area around Bardufoss is second to none. It not only provides a harsh Arctic environment, but also a superb mountain flying area with peaks rising to 5,000 feet, steep-sided valleys and ideal low-level navigation areas, below and above the treeline. The area covers the nearby coastal region with its associated fjords and micro-climatic conditions to the high, wind-swept plains and tundra regions inland. This region of northern Norway is unpopulated, allowing operational flying training, including NVG flying, to be undertaken at any time. During the winter months Bardufoss also boasts some of the worst operational

A Commando Sea King on exercise in Northern Norway picking up a Norwegian Army Recce team from a mountain near Tromsø. (Patrick Allen)

Minus 30 degrees C and 845/846 NAS Sea Kings return to their FOB site during the final Clockwork FOBEX in the winter of 1998. (Patrick Allen)

conditions imaginable where students not only have to cope with flying and fighting effectively, but also with survival in the harshest of Arctic environments.

Flying and operating in winter is a matter of applying known techniques to overcome a potentially hostile environment, and everything at Clockwork is geared to this end. Those undertaking the training programme know that what they experience at Clockwork will probably be the most

difficult conditions they will ever have to cope with. If they can operate effectively here, they can do so anywhere.

The Clockwork training programme

There are two Clockwork training courses each winter: Clockwork 1 in January is dedicated to 846 NAS, and Clockwork 2 in February to 845 NAS. Normally there will be five Westland Commando Sea King HC4s plus ground and aircrew comprising both *ab initio* personnel and those requiring refamiliarisation in Arctic operations.

During their Clockwork training, squadron personnel gain experience of all the problems associated with cold-weather operations, with maintenance personnel having to work on their aircraft out in the open. Once at Bardufoss, the Sea Kings are 'cold soaked' and remain outside permanently. This provides the aircraft engineers and maintainers with experience in looking after and servicing their aircraft in Arctic conditions, which includes engine and transmission changes out in the open in temperatures as low as -30°C.

They must also learn how to survive in the Arctic without getting frostbite or other damage, and must be prepared to fight to defend their aircraft from a tactical operating base (Eagle Base). During their course they learn how to protect their aircraft from the environment using covers etc. and how to prepare them for flying in severe cold-weather conditions. When temperatures drop to below -20°C, Viking heaters have to be used to preheat engines and gearboxes prior to flying. For much of the time the only protection the ground crews get from the environment is from old parachute canopies used to cover the aircraft.

ABOVE: *The reason the Navy take their Arctic flying training so seriously. The environment is demanding and accidents happen – thankfully very rarely.* (Patrick Allen)

BELOW: *An Arctic striped Commando Sea King picking up a troop of Royal Marines.* (Patrick Allen)

Aircrew training comprises a series of lectures and briefings on local area operations and all the pitfalls of operating in the Arctic. Around twenty flying hours are dedicated to general Arctic flying training, which includes six dedicated to NVG operations below and above the treeline. A further eight hours are assigned to the FOBEX.

The training course covers a full range of subjects from aircraft icing limitations to Arctic troops drills, weight restrictions and basic Arctic flying skills. It also covers snow-landing and load-lifting techniques in recirculating snow and the finer points of Arctic mountain flying by day and night. Most front-line pilots will already be familiar with mountain flying having learnt basic skills during their Commando Sea King training with 848 NAS, or by operating with one of the front-line squadrons. Initial training usually takes place in the Austrian/Bavarian Alps, although regulations in these regions forbid landing on the mountains. In Norway, however, there are no such restrictions and this is often the first opportunity pilots have to land on peaks and undertake simulated troop insertions in mountain regions above the treeline by day and night using NVGs. Pilots are shown how to pick out suitable Hover Reference Marks (HRMs) using their instruments and Radalt, windfinding, choosing and reconnoitring safe landing sites, and the importance of escape routes. Great emphasis is placed on teaching pilots to operate safely and how to make the correct captaincy decisions when under pressure.

RIGHT: *Arctic troop drills. A Sea King lands with the starboard wheel next to the troops who huddle down for protection.* (Patrick Allen)

BELOW: *Rugged mountains near Bardufoss provide a stunning, though potentially deadly, backdrop during a two-ship formation sortie.* (Patrick Allen)

Norway not only provides Arctic cold, but also a wide variation of weather conditions, all within a short space of time and distance. These can quickly catch out the unwary and unprepared, and a great deal of care and time is placed on proper flight and mission planning and choosing routes to complete missions safely and successfully. During even the shortest transit, weather can change from clear skies and low temperatures on take-off to a quick thaw, low cloud, then rain, and there is always the ever-present threat of icing. When transiting through mountain valleys in increasingly poor conditions, it is easy just to 'push on' and find yourself being forced ever higher towards low cloud.

Clockwork helps to show pilots the limitations of their own capabilities and those of the aircraft. They learn all the techniques gained over many years of operating in the region to help overcome these problems. There may be a good tactical reason for flying more difficult routes, such as when deploying the BPG or other Special Forces high into U-shaped valleys, or high above the treeline, in poor visibility or low-light conditions. Pilots are shown how to fly these types of missions safely, and if they feel there will be insufficient outside references, or the weather is unsuitable, they should either not start the mission or recognise when to terminate it and find an alternative. The training officers demonstrate these conditions and procedures during the air exercises and explain the various techniques involved, such as keeping three

LEFT: *A Commando Sea King seen flying through the mountains near Bardufoss on a Clockwork training mission.* (Patrick Allen)

BELOW: *Norwegian troops huddle down as a Sea King lifts away. Troops never move out, or into, the rotor disc as helicopters can settle into the snow, lowering the rotor disc clearance.* (Patrick Allen)

outside references ahead at all times and making sure that the valley side is in view at all times, so if a turn or landing becomes inevitable there will always be sufficient outside references to reduce the risk of disorientation and white-out. They are also shown how to incorporate essential flight instruments into their visual scan on an otherwise VFR flight, so as always to have a reciprocal heading should they lose visual references and have to revert to instruments, rolling on thirty degrees of bank and carrying out a level turn on instruments onto the reciprocal heading.

Crew co-ordination is also an important part of the training, and cancelling or aborting a mission when under pressure to complete it is often the most difficult decision an aircraft captain will have to make. This is particularly true if the mission is already underway and he/she then decides to terminate the mission and turn back, or even to land when he/she feels the weather conditions have become too severe.

More and more emphasis is being placed by military helicopter units on Night Vision Goggle (NVG) operations and the RN commando squadrons, already highly experienced in these operations, lead the way in Arctic NVG flying. RN Commando Sea King pilots receive their initial NVG training with 848 NAS undertaking their Night Owl NVG training package. They must also complete at least two hours' NVG flying every month to remain current. Most front-line pilots fly many more hours than these as they undertake their operational duties in Bosnia or Northern Ireland.

The majority of training at Clockwork both day and night concentrates on operating in heavy recirculating snow and in white-out conditions. At night, training also includes both black and white light operations. Pilots undertake their initial training using conventional night-flying methods which include low-level navigation exercises, circuits, landings and take-offs, and load-lifting from a standard NATO 'T' light. They are shown how to use outside references to reduce the risk of becoming disorientated in the recirculating snow cloud and the use of multiple HRMs during load-lifting. These reference marks can be anything suitable from a tree to a human reference, or a vehicle placed in the one o'clock position. During snow landings, pilots learn to perform zero-zero landings arriving at the landing point with no height and no speed. This keeps the snow cloud created by the rotor downwash well behind the helicopter until the last possible moment. During load-lifting, the helicopter needs to hover above the load as it is hooked on, which creates a huge cloud of recirculating snow. A BV206 vehicle has proved ideal as an HRM, both above and below the treeline. During NVG operations loads or landing sites are marked by an infra-red NATO 'T', or more usually a single Cyalume or IR 'Firefly' light which is invisible to the naked eye. During their NVG flying training pilots are shown the increased dangers represented by wires and pylons, as both can easily merge into the monochrome background. The Sea Kings in recent years have been fitted with a number of IR lights, including the Grimes spotlights, Brightstar IR floodlight and NVG formation lights and anti-collision lights.

Once aircrews have mastered the basic art of NVG route planning, low-level navigation and coping with recirculating snow below the treeline, they begin flying missions above the treeline. This provides them with some of their most challenging flying, and operations in this area become much more critical. Above the treeline usually involves operating high in the mountains or valleys. In this snowscape scenery, particularly when viewed through NVGs in poor ambient light conditions with little shadow and contrast, it becomes very difficult to establish depth perception. Pilots are shown how to maintain their outside visual references at all times, even during the transit, and to watch out for rising ground merging with the sky. During transits and landings over this snowscape pilots must maintain multiple outside references at all times with the second pilot and aircrewman providing a running commentary, calling out heights and speeds and the position of any recirculating snow cloud, while also keeping a good lookout.

With the ever-present risk of icing there are no IMC helicopter operations in Norway. One of the first rules the Commando Sea King pilots learn at Clockwork is 'Clear of Cloud in Sight of the Surface' to avoid the risk of icing. When operating in the mountains on NVGs in poor weather and low cloud there is the constant risk of entering cloud. NVGs keep working even in the poorest of conditions, and can give the wearer a false sense of security. They initially overcome poor weather and even snow conditions until they suddenly 'shut down'. Pilots are taught to look out for early signs of deteriorating weather and to be cautious when taking off into a grey and featureless sky.

For the crewman in the back, there is the added novelty of working with Arctic-equipped troops and their heavy equipment, which includes pulks, skis and Bergens. The average weight of an Arctic trooper is 320 lb plus a pulk, this can average 2,560 lb for eight troops plus the pulk. Aircrew need to be aware that when troops are huddled in the helicopter down-draught, they can easily suffer frostbite due to the windchill created by the rotor downwash. There is also the added risk of the rotor disc being closer to the surface as the helicopter settles into the snow, especially if there is rising ground nearby. Arctic troops are taught to huddle down and the helicopter will land with its starboard front wheel and main cargo door next to them. When troops are being moved, the aircraft heating is turned off so as not to overheat the troops in their heavy clothing or make cold weapons sweat, which will then freeze on disembarking.

The final FOBEX sees aircraft and personnel deployed to a FOB for a week of tactical exercises where they can put all their new-found knowledge to the test. This allows aircrew and maintainers to live and operate effectively, day and night, from a dispersed flying site (Eagle Base) and to undertake operational missions. During the exercise the FOB will be moved at least twice to provide a realistic scenario.

The Royal Navy believes that there is no black art about operating successfully in the Arctic. Clockwork is dedicated to teaching RN Commando squadrons how to adapt and apply known techniques to overcome a potentially hostile environment. This not only provides the squadron personnel with a great sense of professional achievement, but has also proved to be invaluable background for operations elsewhere. Having completed the course, squadron pilots are well prepared and trained to undertake the most demanding of missions, including the covert insertion of advance forces at night using NVGs – wherever the customers want to go.

DUTCH
AIRMOBILITY

In September 1996, just nine months since the first of an eventual thirteen Boeing International CH-47D Chinooks arrived in the Netherlands, No. 298 Squadron, Koninklijke Luchtmacht KLu (Royal Netherlands Air Force/RNLAF), based at Vliegbasis Soesterberg, was declared 'Limited Operational Ready' and became available for operational tasking in support of the newly formed and re-equipped Dutch 11th Luchtmobiele Brigade (11th Airmobile Brigade). On 3 May 1996 the first of seventeen new Eurocopter AS532 U2 Cougar Mk 11s arrived with No. 300 Squadron, also based at Soesterberg and beginning their 'in-house' tactical operation conversion training programme on 26 August initially to train thirty-six RNLAF Cougar pilots for their future operational missions. By the end of 1996 No. 301 Squadron, based at Gilze-Rijen, had

taken delivery of twelve leased ex-US Army AH-64A Apaches which will be operated until the arrival of the first of thirty new NAH-64D Longbow Apaches in the autumn of 1998, when No. 302 Squadron will be activated. These new helicopters will join No. 299 Squadron, also based at Gilze-Rijen, which operates a fleet of twenty-four MBB BO-105 CBs operating in the reconnaissance, command and control and light utility roles. This squadron will retain a small flight of ten Alouette 111s for *ad hoc* tasking/aircrew training etc.

In 1991, the Netherlands Defence White Paper proposed the formation of an airmobile brigade to spearhead the new Dutch defence policy for a more mobile and flexible armed force to adapt to the changing national and international

Dutch 11th Airmobile Brigade troops signal they are ready to the Dutch Chinook crewman. (Patrick Allen)

security situation. This was reaffirmed in a further Defence Priorities Review in January 1993 which concluded that an airmobile brigade could and would provide the necessary mobility and flexibility for high- and low-intensity warfare and could also be used for humanitarian relief and peace-keeping. This resulted in the formation of the new Koninklijke Landmacht, 11th Luchtmobiele Brigade (11th Airmobile Brigade), which was duly established in 1993 comprising three full battalions (2,700 airmobile troops) along with

support/logistics units etc. with mobility and armed protection provided by a fleet of attack and light/medium transport helicopters. This would allow the brigade to operate independently or as part of NATO's Multi-National Division Central under the ACE Rapid Reaction Force (ARRC), working alongside British, Belgian and German airmobile brigades or assigned to the United Nations (UN), the Western European Union (WEU), the Organisation for Security and Co-operation in Europe (OSCE), or any other kind of multinational or *ad hoc*

A 298 Sqn Chinook lifting 20mm mortars with a Cougar in the background. (Patrick Allen)

The Dutch operate one of the most advanced Chinooks, with a fully digital 'glass cockpit' and databus. (Patrick Allen)

ABOVE: *The Dutch operate thirty new Boeing AH-64D Apaches and leased twelve AH-64A Apaches prior to the arrival of their new 'D' models.* (Patrick Allen)

ABOVE: *The new Cougars are operated by 300 Squadron at Soesterberg.* (Patrick Allen)

BELOW: *A Dutch Chinook seen with an Apache. The Apache will help protect the new Chinook and Cougar fleets during Airmobile missions.* (Patrick Allen)

The RNLAF is the first military customer for the Eurocopter AS532U2 Mk 11 Cougar which can transport 28 troops or a 3,000 kg (6,400 lb) sling load. (Patrick Allen)

context. The Defence Paper also proposed that the Dutch military should transfer from a conscript army to an all-volunteer professional army with the new 11th Airmobile Brigade spearheading this change. It was also reaffirmed that the KLu/RNLAF would continue to provide helicopter support to the army.

By the spring of 1997 all three new RNLAF helicopter types (Chinook, Cougar, Apache) were operational, undertaking exercises in support of the 11th Airmobile Brigade within Holland, Poland and further afield. By 1998, the RNLAF will be operating one of the most capable and sophisticated helicopter fleets within NATO and will be ready to support their main customers, the newly-formed and re-equipped Koninklijke Landmacht, 11 Luchtmobiele Brigade, and to undertake a multitude of missions both nationally and internationally. By the time the first NAH-64D Apaches enter service the RNLAF will be well prepared for whatever comes their way in the twenty-first century.

THE MARITIME
RAF/RN HARRIER FORCE

uring her 'Ocean Wave 97' deployment, while operating in the Persian Gulf in March 1997 HMS *Illustrious* was involved in two unique events: 'Operation Jural', which saw the Royal Navy's first operational Fighter Air Defence missions in support of 'Operation Southern Watch', the policing of the southern Iraq no-fly zone, and 'Exercise Hot Funnel', the first maritime deployment of a Royal Air Force front-line Harrier GR7 Squadron, No. 1 (Fighter) Squadron, embarked on a Royal Navy carrier. The HMS *Illustrious* Carrier Air Group (CAG) for Ocean Wave 97 comprised six Harrier F/A2s from 801 NAS, seven Sea King HAS6s from 820 NAS, two Sea King HC4s from 846 NAS and three Sea King AEWs from 849 NAS (B Flt). During the Gulf deployment three 820

NAS Sea King HAS6s embarked aboard the new RFA Fleet Replenishment Ship, the RFA *Fort George*. The 801 NAS complement of pilots during the deployment was eight, with one pilot based at the Combined Joint Task Force (CJTF) Headquarters located at Eskan near Riyadh, acting as the RN Air Liaison Officer during the Operation Jural missions.

Operation Jural

The purpose of the Royal Navy Harrier participation in Operation Southern Watch, in which UK contribution was known as Operation Jural – possibly the first fighter (Offensive/Defensive Counter-Air) missions to be flown over Iraq by UK aircraft – was to provide additional operational

RAF groundcrew working on a Harrier GR7 in the unfamiliar surroundings of the HMS Illustrious *hangar deck. (Patrick Allen)*

experience for the Navy Harrier pilots undertaking missions within a JCTF. It also provided experience for the ship, helping to prove future operational concepts within a JRDF of naval air power providing the necessary range, flexibility, mobility and independence of operations without the need to establish fixed land-based airheads.

Exercise Hot Funnel

On 28 February, 1997 four RAF Harrier GR7s belonging to No. 1 (Fighter) Squadron embarked aboard HMS *Illustrious* as she sailed off the coast of Oman. Under the title Exercise Hot Funnel, this was the first deployment of an RAF front line Harrier GR7 squadron aboard a Royal Navy carrier. The

purpose of the carrier deployment was to prove and expand the concepts of RAF Harrier GR7 operations from RN carriers and conducting mixed fighter force operations with RN Sea Harrier FA/2s, and to refine maritime operational capabilities. It also allowed the deck qualification of nine RAF Harrier pilots (eight plus the OC) the introduction of squadron ground and engineering personnel to carrier operations prior to the squadron declaring it was fully operational to undertake maritime duties in support of the UK's new JRDF.

As part of the new JRDF the combination of RN Harrier FA/2s acting in the OCA/DCA role plus Harrier GR7s providing close air support – for example, in support of 3 Commando Brigade, Royal Marines during an amphibious assault – would

An RAF Harrier GR7 launches from the ski ramp of HMS Illustrious *in the Persian Gulf. This method of departure was unfamiliar to most RAF Harrier pilots prior to Exercise Hot Funnel.* (Patrick Allen)

ABOVE LEFT: *An RAF Harrier GR7 recovering to the deck of HMS* Illustrious *during Exercise Hot Funnel.* (Patrick Allen)

ABOVE RIGHT: *A Sea Harrier FA/2 crests the* Illustrious *ski ramp at full power en route to an escort mission over Southern Iraq.* (Patrick Allen)

provide a highly effective force. With their day/night capability, the GR7s would give the brigade much-needed support during the critical night or first-light phase of their amphibious assault missions as they establish their beach-head, with the Harrier force operating from a carrier base until moving ashore to operate in the standard RAF role alongside the troops from the FOB.

The Royal Navy looks upon its carriers as a defence asset. In the JRDF role they are flexible enough to provide a platform for Navy, Army and RAF helicopter or Harrier operations, while also providing all the command and control (C3) systems for national or joint force operations anywhere in the world. The Carrier Air Group (CAG) can be tailored for specific roles. In the support of amphibious operations, for example, the CAG could be aircraft heavy with ten Sea Harrier FA/2s and six Harrier GR7s plus a flight of Sea King AEWs for airborne

command and control etc. Operating from international waters a carrier needs no diplomatic clearances and can loiter or roam freely. A carrier can travel around 670nm in a day, losing itself in around 2,464 square miles of ocean. Capable of rapid deployment and providing a formidable asset the aircraft carrier can deliver a heavy blow from her air or missile assets at very short notice.

The new RAF night-capable Harrier GR7 will provide a formidable addition to the UK's maritime capabilities. The effort and enthusiasm shown by both the RAF air and ground personnel from No. 1 (Fighter) Squadron as they perfect and expand their new operational capability was evident during the visit. As was shown during the Falklands War, the RAF Harrier force was an important addition to the UK's maritime capabilities and one which it would seem a pity not to exploit to the full.

An AMRAAM-armed Sea Harrier FA/2 starts its take-off run down the deck of HMS Illustrious in the Persian Gulf. (Patrick Allen)

An AMRAAM-armed Sea Harrier FA/2 has a final IFF check by the flight deck crew prior to the first Operation Jural mission over Southern Iraq. (Patrick Allen)

ITALIAN NAVY
AMPHIBIOUS SUPPORT

On 3 March 1997 the Italian military undertook a classic services protected evacuation known as non-combatant evacuation operation (NEO) of Italian and other nationals from Albania which included the rescue of German and Dutch reporters being kept hostage in Valona. Valona was the centre of the rebellion against the Albanian government following the collapse of a pyramid saving scheme which in turn led to an armed insurrection and the total collapse of law and order. A number of foreign governments initiated missions to evacuate their nationals including the Americans, who used USMC helicopters to rescue their embassy staff and citizens.

At the same time the British began a contingency plan to reinforce their embassy and arrange the evacuation of their nationals. This initially comprised a combined force of two RAF Chinooks and two 845 NAS Sea Kings based in Croatia as part of SFOR together with an RAF C-130 Hercules which deployed to Goia Del Colle Air Base in southern Italy to fly missions into Albania. The Chinooks, with Special Forces aboard, flew into Albania to reinforce the British Embassy and provide a protection force for an evacuation while the Commando Sea Kings acted as maritime CSAR for the Chinooks with the C-130 Hercules flying top cover and acting as airborne command and control and communications

platform. During the initial phase Chinooks undertook classic decoy landings away from their intended Evacuation Handling and Assembly areas to draw potential troublemakers away and each carried at least twenty-five troops for helicopter protection. This was the first British Combined Force operation to be run by the newly established Permanent Joint Headquarters at Northwood, UK.

The Italian rescue mission, labelled 'Operation Albania', involved CH-47 Chinooks from Reggimento Antares of the Aviazione Leggera Esercito (ALE), plus Agusta SH-3Ds and Agusta Bell 212s from the newly formed 4 Grupelicot, Nucleo Elicotteristico per la Lotta Anfibia (NLA), Marina Militare (Italian Navy). The Agusta SH-3Ds were carrying Special Forces troops from the élite San Marco battalion with two Agusta Bell 212s armed with machine-guns acting as escort gunships. The Agusta SH-3Ds landed their Special Forces troops to seize and secure the proposed landing site for the two Chinooks to pick up the civilians, while the two AB212s circled overhead acting as close air support. Above the helicopter package a pair of 36 Stormo Tornados were flying a protective CAP. The hostages were duly recovered without incident and the Chinooks, Agusta SH-3Ds and AB212s departed for Brindisi in Italy. This was probably the first operational mission undertaken by the Agusta SH-3Ds and

An Italian Navy 4 Grupo Commando Agusta SH-3D used for amphibious operations. (Patrick Allen)

TOP: *Italian Chinooks and 4 Grupo Commando SH-3Ds were operated in Albania during 1997 undertaking NEO missions.* (Patrick Allen)

ABOVE: *The Italian Navy will operate a number of EH101s in the amphibious support role replacing the SH-3Ds.* (Patrick Allen)

BELOW: *An Italian SH-3D from 4 Grupo in formation with an RN Commando Sea King during a joint training mission.* (Patrick Allen)

AB212s from the newly formed 4 Grupelicot. Like other countries within NATO, Italy is looking towards a smaller, flexible and more mobile force structure and has turned to amphibious operations to form part of its Rapid Deployment Force which will be capable of supporting joint Army and Navy operations including out-of-area, peacekeeping and humanitarian missions. This amphibious force, known as the NLAS already has a number of ships including the Landing Platform Dock (LPD)/assault ships the ITS *San Marco* and ITS *San Giorgio* and the light aircraft carrier ITS *Guiseppe Garibaldi* operating Harriers and helicopters and eventually the new EH101 helicopters.

The Marina Militare Italianas (MMI) élite San Marco battalion provided marine commandos for the amphibious force. It is from this unit that Special Forces troops are drawn for the Italian Raggruppamento Sub acqui ed Incursori (Special Sub-aqua Raiding Commando Unit) which undertakes a similar role to the Royal Marine's Special Boat Squadron (SBS). The unit comprises around 200 marine commandos all of whom are highly trained for their specialist roles including skills such as parachuting, swimming and diving.

With their increased interest in amphibious operations, the Italian Navy has ordered an initial four maritime utility EH101 variant helicopters which will provide the airlift for the amphibious troops and which should enter service by the year 2002. In the meantime, modified Agusta/Sikorsky SH-3s are being operated by the recently formed 4 Grupelicot, NLA which provide the specialist aviation support for this new force. Roles include Special Forces, maritime and amphibious operations and fire support for amphibious operations. At the present time 4 Grupelicot is busy expanding their operational roles which include maritime NVG and special operations support and expanding their amphibious capabilities.

Based at Grottaglie Naval Air Station near Taranto, 4 Grupelicot is equipped with modified Agusta/Sikorsky SH-3s and Agusta Bell AB212s. The helicopters have been taken from the existing ASW fleet and have undergone conversion for their new Commando/Special Forces role. Both the AB 212s and SH-3s have been stripped out to remove all their ASW and radar equipment, which has been replaced with troop seating or weapons that has helped reduce their empty weights and increase their lift capabilities. Cockpits have been fitted with NVG-compatible lighting, while a cargo hook, armoured crew seats and composite armour plating have been added to the cockpit and cabin floor areas. Avionics have been updated for their tactical support role. a Doppler navigation system, an AvMAP/Agusta Sistemi Airmaster Mk 11 and an NVG compatible 'knee board' digital moving map updated via a Trimble Centurion GPS, giving a real-time display of the aircraft's position via a stored digital map have all been added.

Regularly training alongside Royal Navy Commando Sea Kings and other specialist NATO amphibious units, 4 Grupelicot is expanding its operational missions in anticipation of the arrival of their new EH101 utility helicopters. The EH101 variant will be equipped with a rear ramp, cargo-handling equipment, load-lifting hook, thermal imaging system (FLIR) and radar. These EH101s will also be equipped with a full self-defence suite armed with 7.62mm/12.7mm machine-guns, and be capable of carrying twenty-four fully equipped San Marco marines or Special Forces.

A 4 Grupo crewman signalling that the 'droop-stops' are in, as an SH-3D shuts down after a training flight. (Patrick Allen)

Italian Airforce Combat SAR Squadrons

Like their navy counterparts the Italian Air Force (Aeronautica Militare Italiana) have been modernising and updating their helicopter fleets for new missions. This now includes Combat Search and Rescue which is the responsibility of No. 15 Stormo with headquarters at Rome-Ciampini airport and comprises four squadrons 82, 83 and 84 Centro SAR and 85 Gruppo SAR based at Trapini, Rimini, Brindisi and Ciampini equipped with the Agusta-Sikorsky HH-3F Pelican.

For their new CSAR role the fleet of AS HH-3F Pelicans have been fully modernised for their role including a new avionic fit. This includes new weather radar, forward-looking infra-red, twin CRTs, VOR/TACAN, LORAN-C, radar altimeter, four-channel autopilot, new radios and NVG compatible lighting. Other additions include radar warning receivers, chaff and flare dispensers and three 5.56mm Minimi machine-guns. The new HH-3Fs regularly undertake Combat SAR training with Italian Special Forces and joint Combat SAR training with other NATO countries. Over the past few years the HH-3Fs of 15 Stormo have deployed to Somalia and Bosnia undertaking their CSAR role.

ABOVE: *An Italian AMI HH-3F Pelican seen during a CSAR training mission in Southern Spain armed with 5.56mm Minimi machine-guns.* (Patrick Allen)

RIGHT: *The AMI have updated and modernised their fleet of HH-3F Pelicans for the CSAR role.* (Patrick Allen)

SPANISH ARMY
AVIATION SUPPORT

The Spanish Army Aviation Force, known as the FAMET (Fuerzas Aeromoviles del Ejercito de Tierra), provides the aviation support for the Spanish Army, which includes the recently formed Fuerza de Accion Rapida (FAR), or Rapid Reaction Force. The FAR comprises units from the élite La Legion Española (Spanish Foreign Legion), Special Forces and the Brigada Paracaidasta (Alcala de Henares), the Spanish parachute brigade.

Colmenar Viejo is the home and principal base of the FAMET. Situated about forty kilometres north of Madrid below the Guadarrama mountains, Colmenar Viejo is also the home of FAMET's headquarters, CEFAMET, the main flying training unit, BTRANS, the signals units and SHEL, FAMET's main helicopter overhaul and third line maintenance unit. It is also

home to Spain's only Chinook squadron, BHELTRA-V, which operates nine modernised Chinook CH-47Ds and a similar number of earlier model Chinook 414/CH 17Cs which support Spanish Army and specialist units throughout Iberia and overseas.

Chinooks, Super Pumas and Bell 212s provide the Spanish Army with their medium/heavy-lift capability with reconnaissance and anti-tank missions undertaken by the BO105s. Chinook missions include logistic support – moving troops, artillery, ammunition and stores, casualty evacuation and outsized cargo, including the recovery of downed aircraft/helicopters. The Chinook can carry thirty-three fully equipped troops sitting on side wall seats, sixty-plus standing and sitting, or twenty-four static-line paratroopers. Internal

All FAMET pilots fly the OH-58, having completed basic flying training. (Patrick Allen)

Bheltra-V Chinooks seen at Colmenar. They operate both CH-47Cs and modernised CH-47Ds. (Patrick Allen)

loads range from several jeep-sized vehicles to three 105mm light howitzers, or bulk loads up to 21,000 lb/9,500 kg. The triple cargo hook system (centre hook 26,000 lb/11,800 kg, fore and aft hooks 17,000 lb/7,700 kg, or 25,000 lb/11,350 kg forward/aft dual point). Chinooks are frequently detached away from their base to support Army units throughout Spain, including La Legion Española based at Malaga and the newly formed Fuerza de Accion Rapida.

Spanish Amphibious Support Operations

As part of the new Fuerza de Accion Rapida, the 1st Spanish Marine Battalion are supported by Spanish Navy Agusta Bell 212s during their amphibious operations. These AB212s embark aboard the Spanish Navy amphibious support ships such as the SNS *Pizarro* and SNS *Aragon* undertaking a variety of roles in support of the Spanish marines battalion.

TOP: *A Spanish Navy AB212 operating in the amphibious support role.* (Patrick Allen)

BOTTOM: *A FAMET CH-47D Chinook picking up a gun in dusty conditions.* (Patrick Allen)

TOP: *The Bheltra-V Chinook flight line at Colmenar Viejo near Madrid.* (Patrick Allen)

ABOVE: *A Spanish CH-47D Chinook working with paratroopers from the Grupo de Artilleria.* (Patrick Allen)

BELOW: *AB212s operated by the FAMET signals battalion BTRANs.* (Patrick Allen)

NORTHERN IRELAND SUPPORT

Over the last twenty-five years counter-terrorist operations throughout Northern Ireland have become increasingly sophisticated as the security forces undertake their daily duties throughout the province. Today, helicopters not only fulfil their more traditional role of providing the security forces with their mobility and airborne protection, but also undertake more specialist roles and multi-agency missions with both ground and air units. For the helicopter crews, Northern Ireland provides some of the most challenging operational flying, particularly at night within both urban and rural environments.

Helicopters are now one of the most important assets in the fight against terrorism in the province, providing the security forces with much-needed mobility, particularly in high risk areas where travelling by road is considered too hazardous. They also provide an airborne surveillance capability, as well as undertaking their normal day-to-day support helicopter roles such as inserting and extracting patrols and resupplying SF bases throughout the province. Other regular tasks include 'Eagle' VCPs, where troops are flown into an area to set up instant checkpoints and then just as quickly relocate to other areas. Patrols and Eagle VCPs inserted by helicopter (day and night) have given security force operations immense flexibility and unpredictability – key words in counter-terrorist operations – leaving the terrorist 'wrong footed' regarding planning and never quite certain as to the disposition or strengths of the security forces. Heliborne operations are particularly valuable in rural areas and have been proved to help deter terrorist activities. Helicopters continue to support the security forces throughout Northern Ireland, whether it's observing the streets of Belfast or deploying security forces in the hills of South Armagh. Until a permanent peace settlement is agreed, they will remain a common sight in the skies.

An RAF Puma seen at Bessbrook with a pair of Army Lynx lifting away for a patrol in South Armagh. (Patrick Allen)

ABOVE: *RN Commando Sea Kings returned to the Province in 1993 after an absence of almost ten years.* (Patrick Allen)

RIGHT: *An RAF Puma armed with a GPMG departs RAF Aldergrove for a day's tasking.* (Patrick Allen)

OPPOSITE TOP: *Army Gazelles fitted with FINCH and Nightsun are used for airborne surveillance.* (Patrick Allen)

OPPOSITE MIDDLE: *An Army patrol returns to Bessbrook after another patrol in bandit country.* (Patrick Allen)

OPPOSITE BOTTOM *Army Lynx are the workhorses in South Armagh. They are fast, agile and armed with door-mounted GPMGs.* (Patrick Allen)

ABOVE: *The RAF Wessex is coming to the end of its operational life in the Province, being replaced by the faster Puma.* (Patrick Allen)

BELOW LEFT: *No. 72 Squadron Wessex with IR jammer.* (Patrick Allen)

BELOW RIGHT: *During high threat periods in the Province, RN Sea Kings are armed with door guns and the crews wear 'body armour'.* (Patrick Allen)

FRANCO-BRITISH EUROPEAN AIR GROUP

Originally formed in September 1995 and inaugurated by President Chirac and Prime Minister John Major on 30 October 1995, the Franco-British (European) Air Group (FB(E)AG) was established to strengthen co-operation between the Royal Air Force and the Armée de L'Air and to promote inter-operability and complementarity between the two air forces in all aspects of air and ground operations except nuclear and NATO Article 5.

Headquartered at RAF High Wycombe, with a permanent staff of four British and four French officers and tasked by a steering group, the FBEAG is sponsored to a series of annual exercises to validate its work. These exercises, involving RAF and French Air Force (FAF) air and ground units, are named 'Volcanex'. The first 'Volcanex 96' took place in September 1996 in the north-east region of France. The exercise scenario was based around the recovery of British and French nationals taken hostage. The second exercise in the series, 'Volcanex 97', took place in the West Country and Wales during June 1997 and was based on a humanitarian aid/peacekeeping scenario and again involved RAF and FAF air

and ground units including air force paratroopers from the No. 2 (Parachute) Squadron, RAF Regiment and the FAF, Commandos de L'Air. All the Volcanex exercises are based around realistic and topical scenarios and are intended to test the procedures and theories developed within Air Group Headquarters, and to rehearse other air and ground procedures being developed nationally.

The initial success of FBEAG resulted in Germany and Italy being invited to join as full members on 1 January 1998 when the FBEAG was renamed the European Air Group (EAG), retaining its headquarters at RAF High Wycombe.

Commandos de L'Air

Originally formed in 1936, the Groupement Fusilier Commandos de L'Air belong to the Armée de L'Air (French Air Force) and are parachute-trained. Their primary role is to provide ground security for air force installations including airfield and forward operating bases etc. Like other élite forces with a specialist capability, the rapid response, versatility and capabilities of the Para commandos provides the French Air

French Commandos de L'Air seen boarding RAF 7 Squadron Chinooks during a Volcanex exercise. (Patrick Allen)

ABOVE: *An RAF 7 Sqn Chinook lifting vehicles during 'Volcanex 96'.* (Patrick Allen)

BELOW: *The 'Volcanex 96' scenario was humanitarian aid. French FAF C-160 Transals flew aid to a FOB which was then moved forward by RAF Chinooks.* (Patrick Allen)

A parachute capability is still important to specialist units such as the French Commandos de L'Air and 2 RAF (Para) Regt. (Patrick Allen)

Force with a unique force which can be used for a wide range of missions including intervention and protection.

No. 2 (Parachute) Squadron, RAF Regiment
No. 2 (Para) Sqn, RAF Regt comprises the only para-trained troops within the RAF Regiment and is also fully mechanised with the Scorpion family of armoured vehicle, or Landrovers etc. Their main tasks are to provide ground air defence of RAF airfields at home and abroad including providing security for RAF Harrier and support helicopter forward operating sites. Like their French counterparts the Commandos de l'Air, they are ideally suited to undertaking a wide range of other roles.

ABOVE: *French Commandos de L'Air troops being moved by RAF Chinooks off the mountains of Wales during 'Volcanex 96'.* (Patrick Allen)

BELOW: *French Commandos de L'Air troops wait to be moved from the FOB site, having been flown into theatre by FAF C-160 Transals.* (Patrick Allen)

French Commandos de L'Air are all parachute trained and can be deployed day or night. (Patrick Allen)

A pair of French Commandos de L'Air watch as another FAF C-160 Transal arrives at RAF Brawdy, the main FOB during 'Volcanex 96'. (Patrick Allen)

AIR GUNNERY
AND COUNTERMEASURES

From the moment they first flew, military helicopters have been fitted with some form of armament to undertake an offensive role. Starting with hand-held automatic weapons, the Korean War saw helicopter armament becoming more advanced with bolt-on machine-guns and unguided rockets. It was during the Vietnam War in the mid-1960s that the US Army moved towards dedicated 'gunships', equipping their Bell UH-1B Hueys with pylon-mounted 'Flexi' machine-guns and Folding Fin Aerial Rockets (FFAR). This eventually led to the Bell AH-1G Huey Cobra, the world's first dedicated attack helicopter.

Troop transport helicopters, in the meantime, were being armed with self-defence weapons, mainly door-mounted machine-guns, leaving the new attack helicopters to provide escort and protection. Over the subsequent years there has been much debate as to whether transport/utility helicopters should be armed to provide them with an offensive role. Suggested weapons have ranged from dedicated turreted heavy cannons to rockets and air-to-air missiles. It has generally been agreed that troop transport/utility/special operations helicopters would continue to carry defensive weapons only, leaving the offensive role to the latest dedicated attack helicopters and close air support aircraft.

Specialist operators such as the USAF Special Operations Force (SOF) MH-53Js, MH-60G Pave Hawks and US Army Special Operations Aviation Regiment (SOAR) helicopters such as the MH-60K and MH-47E are additionally equipped with more powerful 0.50in (12.7mm) GECAI machine guns which can fire 900 rounds per minute of high explosive and armour-piercing rounds, or the 7.62mm M134 Minigun. Larger calibre weapons, and the Minigun in particular, are considered area denial weapons which are capable of heavy and sustained firepower, helping to make heads go down and stay down while a helicopter makes an escape.

In the UK, RAF, Army and Royal Navy, troop transport

An 845 NAS Commando Sea King tests a newly fitted IR decoy flare system prior to arriving in Bosnia in 1992 to undertake UN duties. (Patrick Allen)

helicopters are normally armed with the standard British Army 7.62mm General Purpose Machine-Gun (GPMG). In 1980, the Chinook entered service with the RAF and in 1982 a number were deployed to the Falklands. During this rushed deployment an effort was made to arm RAF Chinooks with the standard British Army 7.62mm GPMG. Gun mounts were hastily welded onto the aircraft, although unfortunately all but one of these Chinooks, along with their newly made GPMG gun mounts, were lost aboard the MV *Atlantic Conveyor* when she was hit by an Argentine Exocet missile. It was not until the build up for the Gulf War in 1990 that the RAF Chinook fleet acquired the M60D and M134 Minigun. It was logical for the RAF to re-equip its Chinook fleet with the M60D and M134 Minigun as they were the standard weapons fitted to US Army CH-47Ds and Special Operations MH-47Es and were already combat-proven in Vietnam and later conflicts in Grenada and Panama.

The Minigun is fitted to US Army Special Operations Chinooks or Black Hawks, and these are equipped with an NVG-compatible laser sighting system which puts a laser spot onto the target area. Over the years the Minigun has proved an ideal area denial weapon capable of placing large amounts of firepower quickly over a large area. The clandestine role of Special Operations helicopters often requires them to deploy long distances into hostile territory, usually unescorted. Weapons such as the Minigun help provide additional protection should they come under attack during the more vulnerable phases of their missions, particularly as they approach or depart a landing site or are waiting on the ground picking up or dropping off troops.

NVG Night Sights
During night operations, RAF Chinook crewmen fire the M60D and M134 Minigun while wearing their Nite Op NVGs. Rounds, in particular tracers, are easily visible through their NVGs although the gunner needs to

TOP: *An RAF Chinook launches flares in Bosnia during an IFOR deployment.* (Patrick Allen)

ABOVE: *IR decoy flares are now fitted to most military fast jets. This RAF F4 Phantom is seen rippling-off flares over the Falkland Islands.* (Patrick Allen)

OPPOSITE: *An 845 NAS Sea King launching forward firing flares in Bosnia. Forward firing flares burn alongside the helicopter for longer before falling away, helping to confuse an incoming missile.* (Patrick Allen)

ABOVE: *US Army Special Operations MH-47E Chinooks have one of the most comprehensive aircraft survivability suites fitted to any helicopter.* (Patrick Allen)

BELOW: *RAF Chinooks are all fitted with full NVG compatible cockpits and full aircraft survivability equipment including Sky Guardian RWR.* (Patrick Allen)

A USAF C-130 Hercules displays its flare firing capabilities. Transport aircraft are now being equipped with infra-red, fine tracking and laser beam jammers. (Lockheed-Martin)

concentrate on the target area adjusting his shots by the tracer, as the gun sights are out of focus. Tracer rounds, however, are extremely bright and can affect the performance of the NVGs 'blowing out' vision as the goggles close down due to the light. To help overcome this limitation, and to increase first-shot hits day or night, a new NVG-compatible sighting system has now entered service with both the RAF and RN commando squadrons to help increase the gunner's chances.

Ammunition manufacturers have seen the need for an

NVG-compatible tracer round; at least one has produced a round for armoured fighting vehicles fitted with night vision systems. Normal tracer rounds are too bright for the latest Generation 111 NVGs with the target area normally obscured by tracer rounds even when used at a 1:8 ratio. New NVG-compatible ammunition helps alleviate this problem, although operationally it will probably require carrying day and night ammunition and will only be viable for Special Operations units, especially when using heavier 0.50inch (12.7mm) calibre weapons.

Aircraft Survivability Equipment (ASE)

Military aircraft and helicopters, particularly Special Operations, have never before faced so many threats while undertaking their missions. As threats increase and become more sophisticated, Aircraft Survivability Equipment and countermeasures also increase in effectiveness. There are numerous pitfalls awaiting the unwary, including radar-guided, IR homing and laser-guided weapons – multiple guns, rockets and missiles. Not only are the threats becoming more sophisticated, but modern shoulder-launched missiles leave little time for aircrew to respond to an engagement as these missiles are long-range, high-velocity (reaching Mach 4) and almost impossible to jam. Laser-guided weapons are also increasing in numbers and capabilities. These all require aircraft and helicopters to be equipped with suitable ASE, and US Army Special Operations helicopters like the MH-47E and MC-130 Talons have full survivability equipment including radar warners, laser warners, missile approach warners, chaff/flares, IR jammers, exhaust suppressors, armoured crew stations and IR reflective paint schemes, all helping to avoid being detected, and if detected allowing the crews to assess the threat and then counter any engagement.

As threats become more sophisticated and reaction times reduce, the military is now turning more towards programmable, automated countermeasures systems capable of spotting the threat then nullifying it automatically. The UK military helicopter fleet, especially the RAF and Commando squadrons, have been quick to realise the value of ASE and their Chinook and Sea King fleets are fully equipped with the latest systems. This will soon include a fully automated missile countermeasures system known as NEMESIS which will not only equip the UK's Support Helicopter Force including the new Merlin HC3 and Chinook HC3, but also the RAF's C-130Js along with US Army and USAF Special Operations aircraft/helicopters.

Like ammunition, infra-red countermeasure flares, fired either manually or automatically linked to the ASE, need to be NVG-compatible, and 'black-light' flares are now entering operational service. For Special Operations aircraft on clandestine missions inside enemy territory, the last thing the crew wants is to see the sky light up with their flares, even if the aircraft has been targeted. Today's RAF support helicopter survivability systems are comprehensive and include Doppler-shift AAR-47 Missile Approach Warners (MAW) linked to an AN/ALE M130/40 flare launcher system which can be fired automatically or manually, the Sky Guardian 2/19 programmable Radar Warning Receiver (RWR) linked to chaff/flare launchers, plus AN/ALQ-157 IR jammers. These systems are continually updated as new technology emerges and will alert the crew to almost any hostile threat. RAF Chinooks are also painted with IR reflective paint to help reduce IR signature.

RIGHT: *A US Army Black Hawk armed with M60 machine-guns seen operating in Northern Iraq.* (Patrick Allen)

OPPOSITE RIGHT: *A USAF Special Operations MH-53J Pave Low fully armed with 7.62mm M134 Minigun, 0.50-inch Gecal machine-guns and IR jammers. These helicopters have full RWR/Laser and missile protection systems and operate almost exclusively at night.* (Patrick Allen)

ARNG
COUNTER-DRUGS SUPPORT
AND EMS

Within the past few years US Army National Guard (ARNG) aviation units located throughout the United States have been working alongside law enforcement agencies in the fight against illegal drug operations. Over thirty-two Reconnaissance And Interdiction Detachments (RAID) have been formed, operating throughout the United States with specially equipped Bell OH-58A1 Kiowas to assist the Drug Enforcement Agency (DEA), police and other law enforcement aviation units in federal and state counter-drug operations. During a single year (1995) ARNG soldiers and aviators participated in 4,182 operations totalling 1,109,359 duty days assisting in the fight against drugs.

The ARNG is seen as the people's army – ready for action both strategically and nationally when called upon by the President and by the governors of the states and territories. Its federal role includes supporting national security objectives as well as deploying and fighting alongside regular army units; its state role includes protecting life and property, preserving peace, order and public safety, and providing a community role giving invaluable assistance during national or state emergencies and crises. Today, this now includes the fight against illegal drug operations throughout the USA.

One such ARNG RAIDs unit is operated by the District of Columbia, based at Davison Army Airfield, Fort Belvoir, Virginia, seventeen miles south-west of Washington DC, the nation's capital and one of America's most violent cities. This RAIDs unit works alongside other Washington law enforcement aviation units including the Metropolitan Police Aviation Unit and the US Park Police Aviation Unit in their battle against drugs.

For their counter-drug missions the ARNG RAIDs Bell OH-58 Kiowas are equipped with specialist equipment. Each RAID has two Bell OH-58A1 aircraft fitted with a FLIR 2000 system for covert surveillance, plus a Wulfsburg radio system

A MEDSTAR BK117 rushing another casualty to the Trauma Center in Washington DC. (Patrick Allen)

A MEDSTAR BK117 launches on another EMS mission in Washington DC. The BK117 is a popular helicopter for EMS missions, capable of carrying two stretcher patients, pilot and two paramedics/doctors. (Patrick Allen)

ABOVE: *The ARNG RAID OH-58A+ helicopters are crewed by two ARNG pilots and a law enforcement observer in the rear.* (Patrick Allen)

OPPOSITE TOP: *Washington Metropolitan Police MHD500Es equipped with Nightsun work closely with the DC ARNG on drug missions.* (Patrick Allen)
OPPOSITE BOTTOM: *A Washington Metropolitan Police MHD500E leaving Washington Executive Airport on another police support mission.* (Patrick Allen)

US Park Police operate Bell 412s, which are used to transport SWAT teams and in SAR/casevac roles. They can be fitted with a FLIR. (Patrick Allen)

which facilitates a multitude of communication possibilities similar to a telephone party line. This system allows up to three different frequency bands to be received and to transmit to one another, as a result the OH-58s can be used as a communications platform relaying information between air or ground units that may normally have incompatible radio systems, or they can undertake an airborne command and control role working alongside fixed-wing aircraft carrying specialist long-range communication systems. The FLIR 2000 system is capable of acquiring and recording images by day and night. The FLIR image can be switched from wide to narrow field-of-view enabling the user to zoom in on a subject without having to adjust focus, gain or contrast. The co-pilot controls the FLIR using a hand-held system controller. Two FLIR monitoring screens are located in the front cockpit, with a FLIR screen and VCR in the rear. ARNG pilots, the majority of whom are ex-regular Army and highly experienced, also wear ANVIS-6 NVGs which, although often degraded by the high ambient light from built-up areas, help to supplement the FLIR imagery in the darker urban areas and parkland that surround the capital. Navigation equipment consists of LORAN, which is supplemented by a GPS.

There are strict rules covering the use of the ARNG helicopters in counter-drug operations, which outline the responsibilities between the Guard detachment and law enforcement agencies. All missions must be drug-related and the Guard detachment is not allowed to initiate any missions. Missions can include aerial/photo reconnaissance, marijuana greenhouse/drug laboratory eradication/detection, transport support, SAR for lost persons and aerial interdiction support using area surveillance and reconnaissance, which can mean the observing and tracking of seagoing vessels, aircraft and ground vehicles suspected of involvement in drug activities. They are, however, not permitted to conduct systematic surveillance deliberately directed at a particular individual and will not become involved in any arrests, searches of individuals or contraband seizures. Guard personnel must remain outside the chain of evidence and custody, and during missions the OH-58s must carry a member of the law enforcement agency who will load and operate the VCR tape recorder should this be needed for evidence. The ARNG is not permitted to store or hold any gathered information.

Capital EMS

Also operating out of Washington DC is a dedicated Emergency Medical Service (EMS) helicopter operator,

ARNG RAIDs operate at night using both FLIR and NVGs. They are equipped with Wulfsburg radio systems to communicate with other law enforcement agencies. (Patrick Allen)

The cockpit of an ARNG RAIDs OH-58A+ showing Wulfsburg radios and twin FLIR screens. The FLIR is controlled by a hand-held system operated by the co-pilot. The police observer loads the VCR tapes as police evidence. (Patrick Allen)

MEDSTAR, which flies two MBB-Kawasaki BK117s and is based at Washington's two major hospitals, the Washington Hospital Center and the Washington Children's Hospital. These hospitals have some of the most advanced trauma, burns and cardiac centres in the world, providing the fastest and best in medical care.

MEDSTAR, short for Medical, Shock, Trauma and Acute Resuscitation, was started in 1983 operating a Eurocopter Bo105 which was replaced by the two BK117s in July 1983. They are operated under lease by Corporate Jets Inc. of Pittsburgh which specialises in providing dedicated EMS helicopters and trained pilots. The MEDSTAR helicopters operate from a purpose-built helipad outside the main Hospital Center along with a second helipad on the roof of the Children's Hospital. They provide full trauma and emergency response and are capable of being airborne within four to five minutes. The helicopters regularly recover road accident victims, gunshot and other casualties from around the Washington area; they also transfer patients to other hospitals as far away as New York. The machines operate day and night and undertake VFR duties only, although the

pilots are all instrument rated should they inadvertently go IMC and have to land at Washington National or Dulles Airport.

The BK117's crew comprises one pilot, one nurse and one paramedic. The machines are fully equipped for medical emergencies and can carry two stretcher patients who are normally loaded through the rear clam-shell doors.

Flying EMS missions over a large city or urban area is demanding, with pilots having to approach and land at unfamiliar landing sites in all weathers, day or night. MEDSTAR pilots have undertaken their own training programme to educate Washington's emergency personnel including firemen, police, rescue personnel, Secret Service etc. in the skills of setting up a suitable emergency landing site along roads and other clear areas, and how to make them secure. They demonstrate how to check out approach and take-off paths and the correct use of radios, hand commands and red flares by day and strobe lights at night to mark the LS. MEDSTAR is typical of many EMS operators throughout the USA which are responsible for the saving of many lives, especially in Washington DC.

GERMANY'S MARITIME BORDER GUARDS

Based at Fuhlendorf, Bad Bramstedt, eighty miles north of Hamburg, Germany, Grenzschutzfliegerstaffel Nord (GSFISt-Nord) provides dedicated aviation support for the German Border Guard (Bundesgrenzschutzes) Northern Command with a headquarters at Bad Bramstedt. The squadron undertakes a wide range of specialist maritime missions and supporting roles including pollution control, the monitoring of shipping within the North and Baltic seas, counter drugs and illegal immigration, which in recent years has become a growth industry as these activities are now the province of large international organised crime syndicates. It also undertakes VVIP missions, assists the local population in times of emergencies such as the floods in the summer of 1997, and provides aviation support for regional police forces

and a variety of more specialist missions supporting both federal and regional authorities and specialist law enforcement agencies including the Coast Guard, Customs & Excise and the Waterways Police. In their more specialist roles, GSFISt-Nord supports federal agencies such as the Bundes Kriminal AMT, formed to fight organised crime, drugs trafficking and kidnapping etc. (similar to the American FBI) and regional police special forces units such as the Specialeinsatzkommando Unit/SEK-Sondereinstzkommando of the Schleswig-Holstein police force. As maritime specialists, GSFISt-Nord also provides helicopter support to the élite German counter-terrorist unit GSG-9 during Maritime Counter-Terrorist (MCT) missions deploying GSG-9 onto oil rigs, ships and other installations both by day and night.

A German Border Guard Puma from GSFISt-Sud, based at Munich, seen operating in the Alps. (Patrick Allen)

GSG-9 is part of the German Border Guard (Grenzschutzgruppe 9) and was formed by Ulrich Wegener after the 5 September 1972 Munich Olympic massacre when eight Palestinians from the Black September group took eleven Israeli athletes hostage, a situation which resulted in the deaths of nine of the hostages and the destruction of a UH-1 helicopter. This disaster prompted Germany and the UK, along with other European nations to form counter-terrorist units.

The main area of responsibility for the Border Guard, Northern Command is Germany's maritime borders in the North and Baltic seas, together with her land borders with Denmark, Netherlands and Poland. As much of its activities involves operating in the North and Baltic seas, GSFISt-Nord, under the command of Polizeioberrat (Lt-Colonel) Harald Krahmuller, is the Border Guard's maritime specialist. For this role the squadron is equipped with eight Bell 212s, four SA330J Pumas and three SA318C Alouette-Astazou helicopters. Two of its Bell 212s are dedicated to the rescue/EMS (Rettungsdienst) role and are based at Eutin, south-east of Kiel (Christoph 12), and Bremen (Christoph 6).

For their maritime role, GSFISt-Nord Bell 212s and SA300J Pumas are fitted with emergency flotation equipment and rescue winches. The Bell 212s were purchased in 1973–4 and are powered by Pratt & Whitney PT6T-3BE TwinPac turboshafts. Due to its maritime role, it is the only unit in the Border Guard equipped with the twin-engined Bell 212s, with other squadrons operating the single-engine Bell UH-1D. The Pumas and Bell 212s are fully IFR equipped and both types can be fitted with a variety of auxiliary fuel tanks to help extend range and endurance.

Both the Pumas and Bell 212s are equipped with the Trimble 2101/0 satellite GPS incorporating a moving map system and FM/VHF and UHF radios with maritime homer. The Pumas are additionally equipped with a nose-mounted Bendix weather/search radar. The six Bell 212s will shortly be equipped with the same system, and subject to agreement with the German Civil Aviation Authority the Pumas will be equipped with suitable NVG-compatible cockpit lighting. GSFISt-Nord also operates a Westinghouse combined TV/FLIR system which can be fitted at short notice onto either the Puma or Bell 212 depending on the mission requirements.

Northern support roles

The work undertaken by the squadron is extremely varied and is one of the reasons why the aircrew find their work and missions so challenging and interesting. Less experienced pilots begin their career flying the Alouette, and those with more experience are qualified to fly both the Puma and Bell 212. The majority of pilots are qualified on all three types.

The main work undertaken by GSFISt-Nord is to police the maritime region and borders of Germany in the North and Baltic seas region. The squadron is primarily tasked with the protection of the frontiers of German territory, and this includes the prevention of illegal immigration and the fight against criminal organisations channelling people and goods across the border. The squadron works closely with all the other maritime and police authorities both within Germany and throughout the Baltic and North Seas region. Combined and joint missions are not uncommon, particularly in counter drugs, human trafficking/illegal immigration and the fight

GSFISt-Nord operate Bell 212s, which can be used to transport specialist élite units including GSG-9 and other Special Forces teams. (Patrick Allen)

A Specialeinsatzkommando squad abseiling from a GSFlSt-Nord Bell 212. (Patrick Allen)

ABOVE: *Schleswig-Holstein Specialeinsatzkommandos wearing Balaclava hoods, being airlifted by a GSFISt-Nord Bell 212.* (Patrick Allen)

OPPOSITE TOP: *GSFISt-Nord EMS Bell 212 (Christoph 12) flying an RTA victim to hospital in Rostock.* (Patrick Allen)

OPPOSITE MIDDLE: *An RTA victim is carried to Christoph 12 having been stabilised by the doctor assigned to the helicopter from Eutin Hospital.* (Patrick Allen)

OPPOSITE BOTTOM: *RTAs are a major task for EMS helicopters. Christoph 12 is seen attending an RTA near Eutin.* (Patrick Allen)

against organised crime. Other priority missions are the prevention and apprehension of those committing maritime pollution with harmful substances in the Baltic and North Seas.

The helicopters regularly undertake night surveillance of shipping movements, and with intelligence gathered from the police and Customs from neighbouring countries the fight against organised crime, smuggling, drug and human trafficking is a priority. The pilots of GSFISt-Nord fly between 260 and 350 hours per year with much of this flown at night on NVGs undertaking maritime/surveillance missions. There are few flying jobs today that offer such a variety of challenging missions and helicopter types.

Helicopter EMS

As well as undertaking operational missions in support of the Border Guard, the pilots and flight engineer/navigators also fly missions in the two EMS/Rettungsdienst Bell 212

helicopters, Christoph 6 at Bremen and Christoph 12 at Eutin.

GSFISt-Nord has provided an EMS helicopter at Bremen since 1973, and at Eutin since 1976. Both units operate to a maximum radius of fifty to seventy kilometres around their area and can provide back-up for neighbouring EMS helicopters should it be required. Based at Eutin Hospital, Christoph 12 began operations initially with a Bo105 before acquiring its Bell 212 in 1978. The majority of its missions are undertaken within a fifty-kilometre radius of Eutin and cover the Kiel and Lübeck area and the coastal region of the Baltic, which is a popular tourist holiday destination in the summer. The two orange-painted EMS Bell 212s are fully equipped to provide the best in primary medical care and have everything the doctor needs for most situations. This includes all the basics such as ECG, pulse oximetry, defibrillator, oxygen, suction, blood pressure ventilation, endotracheal intubation, intravenous fluid/drug administration, chest drains etc. The majority of this equipment, although fitted to the helicopter, has been designed to be quickly removed and is portable. The Bell 212 cabin size allows the comfortable transportation of a single stretcher patient and a sitting casualty along with the doctor and two paramedics, although they normally only operate with a doctor and paramedic. The cabin can also be re-roled to carry up to six stretchers, although this would not be required for their normal EMS operations. The helicopter undertakes both primary casualty and Air Ambulance transportation missions – moving patients from hospital to hospital etc. The helicopters are IFR-equipped and have a 'Skyshout' system, rescue winch and full radio fit including VHF/UHF and Chelton Maritime Homer. The normal crew for the EMS role is two pilots or a pilot and flight engineer/navigator, plus a paramedic and the doctor.

The priority mission of the helicopter is to get the doctor to the casualty as fast as possible. Once at the scene the doctor assesses the casualty, who may already be receiving medical treatment from a paramedic who has arrived by ground ambulance or rescue vehicle, or even from another district. It is the doctor who makes the decision whether the casualty should be airlifted to a hospital, and which hospital this should be – in the Eutin region, for example, there are several hospitals in the immediate area at Kiel, Lübeck, Eutin and Hamburg. The doctor may think the casualty should be flown to a specialist hospital outside the immediate area, or may decide that the casualty should be taken to the nearest hospital by ground ambulance, which is what happens in the majority of cases. Only the most serious cases are flown to hospital.

RIGHT: *GSFISt-Nord regularly practise landing on marine installations such as this offshore platform in the Elbe estuary.* (Patrick Allen)

OPPOSITE RIGHT: *Oil/gas platforms in the Baltic Sea. GSFISt-Nord are tasked with policing the area which includes undertaking maritime counter-terrorist support missions with GSG-9.* (Patrick Allen)

A GSFlSt-Nord Puma winching a crewman onto BG-17, one of the Border Guard's fast patrol boats operating in the Baltic. (Patrick Allen)

ABOVE: *The CO of GSFlSt-Nord flying his Puma, equipped with flotation gear, GPS and Bendix weather/search radar, over the Kiel Canal. (Patrick Allen)*

LEFT: *One of GSFlSt-Nord's tasks is to check pollution in the Baltic and North Sea. They take video evidence and water samples of any incidents. (Patrick Allen)*

UK AIRBORNE LAW ENFORCEMENT

Police aviation has literally taken off in the UK over the past five years and there are now more than twenty-five Police Air Support Units (PASU) operational throughout the country.

Today's police helicopters/aircraft have developed into highly sophisticated machines capable of undertaking a growing number of law enforcement roles which can include patrolling, surveillance, communication-linking and airborne command and control. Standard role equipment now includes the latest developments in high-definition digital TV and thermal imaging systems, helping to provide an increased day and night operational capability to support ground units. Quality video material recorded on these systems has been repeatedly shown to be unchallengeable both in a court of law and for intelligence-gathering purposes. Mission

equipment has been further enhanced by the fitting of encrypted communications systems and the latest solid-state microwave down-links.

Great strides have been made in recent years in the range and quality of low-light television and thermal imaging systems for airborne surveillance. Digital systems provide broadcast quality images day and night. Modern lightweight dual TV/thermal systems provide wide-angle to long-range-zoom stabilised images with autotrackers, and integration of GPS which can be bore-sighted to digital moving map systems to provide the operator with the 'complete' picture, which can then be down-linked to ground units. Down-linking the real-time image from the aircraft's surveillance camera using the latest solid-state systems takes the video and thermal image seen by the camera under the aircraft's nose and converts this

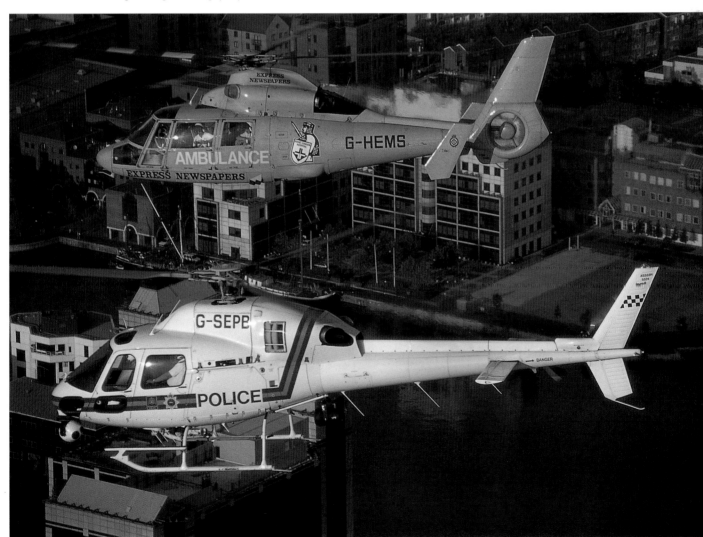

The London HEMS (now Virgin) EMS Dauphin in company with a Met Police AS355N Twin Squirrel. (Patrick Allen)

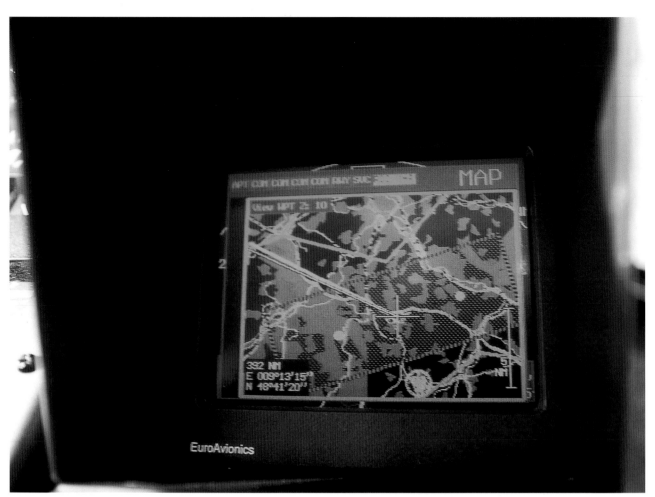

TOP: *The EuroAvionics Euronav 111 GPS digital moving map system is capable of storing the entire Devon & Cornwall police operational area and can be scaled down to show individual properties with the legend showing the position of the helicopter.*

ABOVE: *The latest high-visibility colours for UK police helicopters, helping to reduce the risk of mid-air collisions with low flying fast-jets.* (Patrick Allen)

Helicopters are now an essential asset for a modern police force. (Patrick Allen)

into a microwave signal. This signal, which is encrypted using UK Home Office-approved frequencies, can then be sent to a variety of ground-based receiving stations and directly into the police command and control centres.

In the command and control role the helicopter/aircraft can be fully integrated within established command and control structures allowing senior officers and command teams to have instant access to real-time images from the force helicopter/aircraft, allowing them to deal with an incident as it happens and optimise the decision-making process. The latest technology allows live pictures to be beamed down to portable 'briefcase-sized' colour LCD TV monitors which can be located in a vehicle, for example, or be hand held. Weighing only a couple of kilos, the 'hand-held' will give the 'on-scene' officers first-hand images of the situation allowing hostage negotiators or members of an armed support team that crucial up-to-the-second picture of the scene from the best possible angle.

Devon and Cornwall Police ASU

The Devon and Cornwall Police started helicopter operations as long ago as 1979, beginning with short-term charters of

around six weeks during the busy summer season when the region's population increases by four million holidaymakers, although the use of helicopters/aircraft soon needed to be increased as their cost-effectiveness became apparent. In August 1984, as helicopter operations became more established, a second-hand Eurocopter Squirrel was purchased and the force became an owner/user. In the spring of 1987 the Squirrel was replaced with a brand-new Eurocopter Bo105DBS which was replaced by a larger Eurocopter BK-117C1 in January 1998.

Over the years, the Devon and Cornwall PASU helicopters have been fitted out with the very latest in mission equipment in a high-tech two-crew cockpit helping to increase operational effectiveness.

The Devon and Cornwall helicopter has been equipped with a high-tech two-crew integrated cockpit with the latest in digital mapping, surveillance, communications and mission management technology to provide the force with a multi-mission-capable airborne platform for surveillance, searches, command and control and the many and varied taskings required in the force area. Mission equipment includes the latest BSS4000 high-performance FLIR system 'SAFIRE' digital

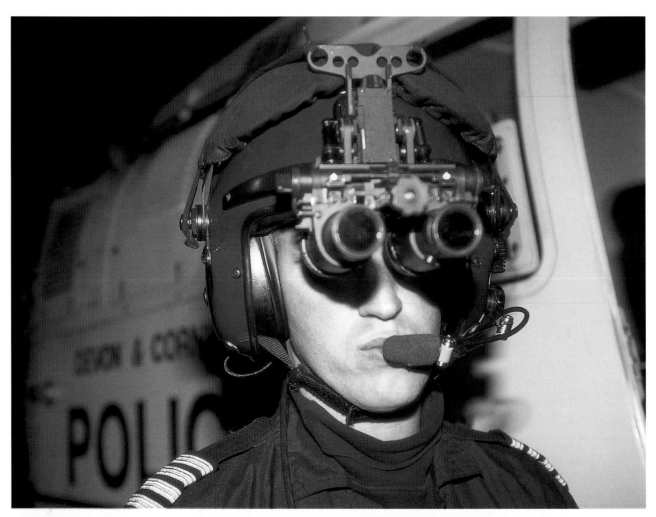

ABOVE: *The comprehensive mission fit on the Devon and Cornwall Police Bo105 has now been transferred to a new, larger BK117.*

OPPOSITE TOP: *The Devon and Cornwall Police ASU have been undertaking trials in the use of NVGs for certain police missions.* (Patrick Allen)

OPPOSITE BOTTOM: *The NVG and night-equipped Devon and Cornwall Police Bo105 seen launching on a night support mission over Devon.* (Patrick Allen)

The Met Police AS355N Squirrels are fully equipped with TV/FLIRs and data down-links for their busy missions over London. (Patrick Allen)

thermal imager and Sony high-resolution video camera with a long-range zoom lens, a Night Sun SX16 searchlight which is slaved to the TV/FLIR camera system, twin 'Skyshout' loudspeakers, Garman GPS receiver, VHF/UHF radios, tracker system and EuroNav 111 GPS-driven digital moving map system, which has been integrated into and bore-sighted to the TV/FLIR display. Role equipment also includes NVG-compatible cockpit lighting.

The EuroNav 111 GPS-driven digital moving map system has helped increase operational effectiveness and reduced pilot/observer workload. This has been further enhanced with the system being linked into the observer's TV/FLIR screen. Displayed on both the observer's TV/FLIR screen and an LCD monitor screen, the moving map system covers the entire operational area and can be scaled up and down to show individual roads and buildings etc.

NVG trial
Operating in conjunction with and closely monitored by the CAA, the Devon and Cornwall Police ASU has been undertaking a two-year operational evaluation of NVG operations. This three-phase trial is hoping to establish that NVGs not only help enhance night flight safety, but also operational capabilities, and will complement other night aids such as the thermal imaging system and LLTV. Unlike the military, which routinely uses NVGs to enhance its night tactical capabilities, NVGs in police operations are seen as an additional aid to assist pilots during transits, increase situational awareness, ease navigation and provide early warning of closing-in weather such as a lowering cloud base. Operationally they will help enhance the observer's capabilities during ground searches in remote areas and provide greater all-round situational awareness. At the present time the only person aboard a police helicopter who operates without the aid of any night sensors is the pilot. The observer sitting in the left-hand seat can use the helicopter's thermal imaging systems to provide an overall view while the pilot is flying on instruments and using any available ground light sources as a hover reference, for example, during surveillance or searches.

UK FIRE BRIGADE AIR SUPPORT

Para-public helicopter operations within the UK continue to expand as the capabilities and versatility of today's modern helicopters are exploited to the full. Having recently established themselves in the Police Air Support and Air Ambulance/EMS roles, they are now being considered by a number of UK fire brigades to help enhance their operational effectiveness.

Over the past few years four UK fire brigades – Strathclyde, West Sussex, London and Mid and West Wales – have been involved in a series of helicopter trials using the Eurocopter BK117. Spearheaded by McAlpine Helicopters Limited, the main distributor for Eurocopter in the UK and Ireland, a proof-of-principle trial was undertaken by the Strathclyde Fire Brigade in the summer of 1994. This proved a success and

was followed by full-scale trials with the West Sussex Fire Brigade in the summer of 1995 and the London Fire Brigade in 1995–6, followed by a feasibility trial with the Mid and West Wales Fire Brigade in autumn 1996.

The trials provided all the agencies involved with a huge amount of experience in this new and developing field. The fire brigades which have been involved are enthusiastic about the increased operational capabilities the helicopters help to provide, and like their police and air ambulance counterparts the role of the helicopter in the specialist para-public support role will continue to increase.

The London Fire Brigade used a BK117C1 in 1996 to conduct proof-of-concept trials, using a helicopter to help increase response times for fires, chemical spills and rescue missions. (Patrick Allen)

ABOVE: *'Some Like it Hot.' The McAlpine Helicopter Ltd demonstrator BK117 seen during trials with the West Sussex Fire Brigade.* (Patrick Allen)

OPPOSITE TOP: *The BK117 delivering a four-man team with specialist clothing and equipment to help contain a chemical spill.* (Patrick Allen)

OPPOSITE BOTTOM: *The BK117 can carry a nine-man team of fire-fighters equipped for specialist tasks.* (Patrick Allen)

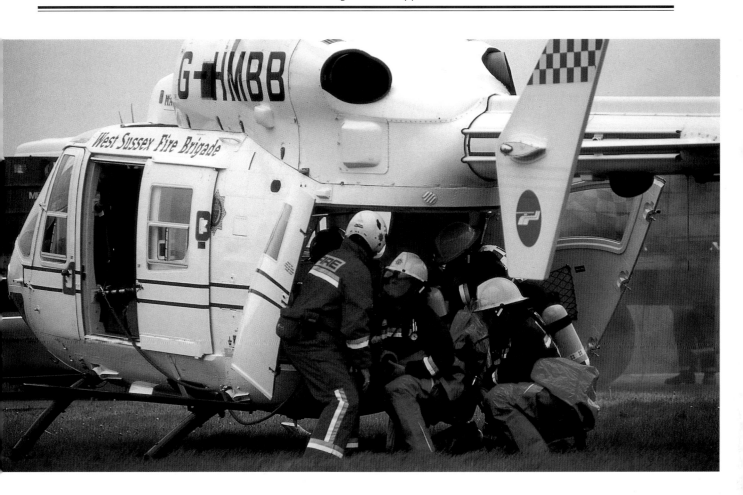

UK COASTGUARD SAR

Like other para-public helicopter operators, HM Coastguard/Bristow Helicopters Limited are making the best use of modern technology to help increase their operational effectiveness while undertaking demanding missions both day and night. The first UK SAR helicopter operator to take advantage of FLIR systems to enhance night search capabilities, HM Coastguard/Bristow Helicopters Ltd has recently updated its aircraft to include the latest TV/thermal imaging systems and integrated navigation and radar systems to help increase mission effectiveness.

The first updated HM Coastguard/Bristow all-weather SAR Sikorsky S-61N began operations at Portland, Dorset in the summer of 1997 with the callsign 'Rescue Bravo-Whiskey'. Fleetwide mission equipment updates include the installation of the Broadcast & Surveillance Systems (BSS) Limited Ultra 4000 dual TV/Thermal Imaging System, replacing the FLIR 2000 system, Racal's RNAV2 Doppler/GPS Navigation Computer, a rear cabin-mounted Trimble GPS and a new lightweight, high-powered landing/search-light. Further updates will include a new cabin layout and a glass Electronic Horizontal Situation Indicator (EHSI). The helicopters are already equipped with Bristow's Health & Usage Monitoring (HUMS), a nose-mounted weather/search radar and full auto-hover/overfly automatic flight control systems.

OPPOSITE: *HM Coastguard S-61s are among the best equipped SAR helicopters in the world.* (Patrick Allen)

BELOW: *A Portland-based HM Coast Guard/Bristow S-61 crewman gets ready to be winched-down to a boat owner in distress.* (Patrick Allen)

ABOVE: *Once safely aboard the helicopter the crew of Rescue Bravo-Whiskey try to make the casualty as comfortable as possible as they head for the nearest hospital.* (Patrick Allen)

OPPOSITE: *The latest TV/FLIR systems are now fitted to Bristow S-61s contracted to HM Coastguard.* (Patrick Allen)

SAR helicopters often work with the UK's highly respected volunteer lifeboat crews. (Patrick Allen)

SPECIAL OPERATIONS CAPABLE AIRCRAFT

Most of the world's military special operations units are supported by aircraft and helicopters taken from everyday operational squadrons which then have specialist role equipment added should this be required. They are, however, usually flown by the more experienced crews who regularly train with the specialist units they support.

Taking aircraft and helicopters from the normal day-to-day operational fleets does have the advantage of giving special operations flight crews the widest possible choice of available aircraft. They can then select the most suitable aircraft, usually the ones with the maximum time between overhaul. Many squadrons maintain a number of aircraft and suitably qualified aircrew on special operations stand-by. Standard aircraft can be quickly re-roled using modular role equipment such as extended-range tanks and fast-roping frames etc. Using everyday aircraft can also help avoid special operations aircraft being recognised for what they are.

Night vision, thermal imaging, satellite navigation /communications and digital mapping systems etc., which only a few years ago were confined to a few American special operations aircraft, are now in everyday use both within the military and para-public units such as police, and the majority of military aircrew are now fully trained in night operations. With today's emphasis on global rapid deployment and more flexible and mobile forces, front-line military aircrew are now being trained to operate in every conceivable theatre of operation including maritime, amphibious and airborne roles. They also regularly train in the more demanding Arctic, desert and jungle environments, missions which were previously only undertaken by Special Forces. Also, many

of the tactics and techniques devised and first used by special operations units are now being taken up by front-line aviation units which are now required to support the new rapid deployment forces, airborne, amphibious and airmobile brigades as they continue to expand their various operational roles.

Although Special Operations Aviation has lost its monopoly on night operations, it is continuing to work at the leading edge of aviation support with regard to tactics, techniques and mission capabilities as they become more and more specialised. Speed, reach, reliability and capability are the key words for Special Operations Aviation, and the only country to develop dedicated all-weather/long-range special operations aircraft and helicopters is the United States of America, to be joined by the British who have acquired a number of dedicated Chinooks.

United Kingdom
Agusta A109
Four Agusta A109As, two taken from the Argentinians after the Falklands War and two purchased new from Agusta and fitted with NVG-compatible cockpit lighting, are used in the liaison and special tasking roles by the military and to assist the civil powers. They are assigned to 8 Flight AAC and should shortly be replaced.

Bell 212/412HP
The British Army operates a number of leased Bell 212HPs which fly in Brunei supporting jungle training, special tasking and the resident infantry battalion. The Bell 412 is also operated by the new Defence Helicopter Flying School and may prove to be popular for specialist roles within the British Army. Both the Bell 212 and 412 have NVG-compatible cockpit lighting.

Westland Lynx Mk 7
Operated by the British Army Air Corps in the light utility and special operations role,

A 7 Flt AAC Bell 212 operating in the jungles of Brunei. (Patrick Allen)

the Lynx can be equipped with a variety of weapons including door-mounted GPMG and 12.7mm machine-guns (FN/M3M or Gecal) and can carry up to ten troops. All Lynxes have NVG-compatible cockpit lighting and some are equipped with GPS/Doppler TANS2 and encrypted radios.

Boeing Chinook HC2

The RAF operates a large number of Boeing Chinook HC2s which can be quickly re-roled with specialist equipment. A number of Chinooks have been additionally equipped with the PRC112 covert Personnel Locator System for the CSAR role. RAF Chinooks have a large range of specialist role equipment available to them including Robertson ERTs and Forward Area Refuelling Equipment (FARE), winches and fast-roping frames, and M134 Miniguns, all of which can be quickly installed depending on the proposed mission. The basic RAF Chinook is equipped to a very high standard in terms of avionics, navigation, communication and aircraft survivability systems, and all have NVG-compatible cockpit lighting etc.

Boeing Chinook HC3

The British Special Forces have purchased eight Chinook HC3s and they are the first dedicated Special Operations helicopters purchased by the British Ministry of Defence and will be operated by the RAF. The Chinook HC3 airframe is based around the US Army's MH-47E and features two 1,030-gallon, long-range, side-fairing fuel tanks, which is double the standard fuel capacity of the Chinook HC2. The HC3 is also equipped with an air-to-air refuelling probe, additional weapon mounts and provision for fast roping and an internal cargo handling system. The two additional long-range tanks give the Chinook HC3 a standard fuel capacity of over 2,060 gallons which allows the HC3 an endurance of five and a half hours. Range can be further increased with the fitting of up to three internal 800-gallon ERTs, each of which extends endurance by about two hours. The HC3 will be equipped with a full 'glass' cockpit with multifunction displays, twin databus with digital navigation and mission management systems with INS/GPS/Doppler navigation systems which are integrated with other mission and operational equipment such as FLIR, weather radar, digital moving map, Satcom and full aircraft survivability systems including radar-absorbing/IR reflective paint. All these integrated systems will allow the HC3 to operate in the worst of weather behind enemy lines.

EH101 Merlin HC3

The RAF has ordered twenty-two EH101 Merlin HC3s which will be based at RAF Benson and will operate in the support helicopter role. This all-weather, long-range helicopter can carry fuel reserves and full role equipment necessary for operational readiness and still be capable of deploying 1,020 km (635 miles) on standard fuel with 1,250 kg (2,750 lb) of spares and support equipment, or 1,300 km (810 miles) with an internal auxiliary tank. With an AAR refuelling capability provided by a detachable refuelling probe which can be fitted in fewer than four manhours, range is unlimited. Operational capability and range is further enhanced by the EH101's de-icing systems which will allow operations in severe conditions. Equipment for the Merlin HC3 reflects the roles envisaged by the RAF in the next century which may also

include specialist support tasks such as Combat SAR, although it will not be as operationally capable as the Chinook regarding versatility or lift capability. The helicopter, however, has a high specification including a full NVG-compatible cockpit with an advanced integrated navigation suite which includes a databus-linked INS/GPS Doppler system along with a multifunction digital mission management/tactical/ navigational system and full defensive aids suite. A dual-duplex automatic flight control system allowing automatic transition to the hover and a FLIR will permit operations in the worst weather. Operational availability is further enhanced by the Merlin being designed to fly over 200 hours between (preventative) maintenance giving fixed-wing standards of reliability. The Royal Navy has purchased the EH101 Merlin as its new maritime helicopter and Royal Navy Commando Aviation is looking to replace its Westland Commando Sea King HC4s within the next few years. A Commando variant of the EH101 Merlin with a rear-ramp is a possible replacement.

Commando Merlin

The Royal Navy's fleet of Westland Commando HC4s will need to be replaced by 2010. These helicopters provide the airborne lift to 3 Commando Brigade, Royal Marines and specialist maritime and amphibious units. The EH101 Merlin helicopter is the obvious replacement, both politically and with regards to operational support. All EH101s are marinised and the Commando variant, known as the Future Amphibious Support Helicopter (FASH), would be equipped with a rear ramp but would also have full automatic tail and blade fold capability. A fully NVG-compatible cockpit based around a MIL-STD-1553b databus and six high-definition full colour displays with flexible mission display options makes for reduced pilot workload and an all-weather operational capability. The digital automatic flight control system incorporating dual-duplex architecture provides automatic stabilisation and autopilot facilities allowing single-pilot operation when tactically appropriate in both VFR and IFR conditions. The AFCS also includes auto-transition to the hover and the GPS/INS navigation system can be interfaced with a digital moving map and FLIR. A real-time data-link will provide digital battlefield/threat/avoidance and safe fly zone data along with real-time information from satellites, AWACS or other forward reconnaissance assets such as UAVs and Special Forces. Survivability will be enhanced by high agility and speed and redundancy of vital systems and structures together with low noise signature. Mission enhancements will include the multi-band ARI 18246 Nemesis Directional Infra-Red Countermeasures System (DIRCM) which will be equipping all US Special Operations and British aircraft/helicopters into the next century. Infra-red countermeasures will also include 'black flares' autocoupled to the Doppler-shift Missile Approach Warners (MAW) to prevent the premature launching of flares/chaff, laser protection, hostile fire detector systems and programmable radar warners. Defensive weapons would include 0.50in calibre machine-guns such as the FN M3M or GE- GUA-19 Minigun. The Commando EH101 would be able to transit 390 miles (630 km) and lift twenty-four troops (287 lb/130 kg each) at high temperatures and out of ground effect over a 125-mile (200 km) radius of action. It could move thirty troops over a radius of 100nm (160 km) or lift internal freight weighing 9,600 lb (4,350 kg) over a radius of forty-seven miles (75 km) or an underslung load of up to 11,103 lb (5,000 kg).

RAF C-130J Hercules

The versatility and operational capability of the C-130 Hercules makes it, without doubt, one of the most important air assets in any special operations role. The RAF originally purchased a fleet of sixty-six Hercules C1s and modified thirty to the stretched C3 variant equivalent to the civil L-100-30 version. All RAF C-130s were equipped with refuelling probes and IR jammers, radar warning receivers and MAWs; a number are equipped with NVG-compatible cockpit lighting. The RAF is now purchasing the updated Hercules C-130J variant due for delivery soon. These are equipped with new 'glass' cockpit with NVG-compatible multifunction liquid crystal displays and two head-up displays, twin autopilots and Allison AE2100D3 engines, new propeller blades and digital engine controls. The new six-blade propellers have increased

thrust by 18% allowing the new C-130J to climb at maximum gross weights to 20,000 feet in just fourteen minutes, instead of the almost thirty minutes in today's C-130. The Allison AE2100D3 twin-spool engines allow for greater operating temperatures, higher altitudes and times between overhaul of 5,000 hours, five times better than the present engines. A MIL-STD 1553 databus has eliminated 600 lb of hard wiring and allowed the integration of the aircraft's systems, keeping critical systems in constant communication with one another. Mission planning from terrain to weather conditions to precise drop sites is now loaded into the C-130J's mission computer via a two-by-three-inch card. The loadmaster has a remote control to allow him to drop cargo more accurately and the rear cargo area is now fitted with NVG-compatible lighting. The new twin head-up displays allow the two pilots to focus outside the aircraft while key instrument readings and navigation information is displayed in front of them. A number of C-130Js will be equipped with Sierra AN/APN-240B station-keeping equipment which automatically keeps each

EH101 Merlins have been ordered by the RN and RAF and a new Commando variant is likely. (Westland)

USAF C-130 Combat Talons are used for night Special Forces support missions. (Patrick Allen)

aircraft in formation position during multi-aircraft airdrops or paradrops. The C-130s will also be equipped with ground collision avoidance and traffic alert systems plus a full defensive aids suite which will include Nemesis. With RAF helicopters (Chinook HC3/Merlin HC3) obtaining an AAR refuelling capability a number of C-130Js will be additionally equipped with underwing refuelling pods. The RAF has ordered two versions of the standard length C-130, designated the Hercules C5, and the stretched version, which is fifteen feet longer, designated the Hercules C4.

WAH-64 Apache

The British Army Air Corps has purchased sixty-seven WAH-64D Longbow Apaches which will be built under licence by Westland Helicopters Ltd. Similar to the US Army's AH-64D Longbow Apache but with British Rolls-Royce RTM322 engines, weapons and avionics, the WAH-64D will provide the Army Air Corps with its first dedicated attack helicopter capable of deep strike operations behind enemy lines. The AAC is busy studying and developing night deep strike offensive operations using the WAH-64's long-range targeting and weapons strike capability to the fullest. It is probable that the corps will group its attack helicopters into three regiments of two AH squadrons each, plus a Lynx Light Utility Helicopter (LUH) squadron per regiment. For their long-range missions

the AAC WAH-64s will require FARPing from RAF Chinooks undertaking a similar role to the deep strike operations undertaken by the US Army's 101st Airborne Division (Air Assault) which obtains its strike range by forward deploying AH-64s to holding areas behind the FLOT where Chinooks provide Forward Area Refuelling and Rearming (FARR) for the Apaches before heading off for their attack targets. The WAH-64 will allow the AAC to conduct manoeuvre warfare for the first time, give the corps the reach to attack targets at greater stand-off distances than ever before and provide the ground army with a true all-weather day/night attack helicopter to fight on the battlefield of the twenty-first century. It will also allow the AAC to integrate its WAH-64s into the new digital battlefield down-linking information from UAVs etc.

Boeing E-3D Sentry

NATO Airborne Early Warning Force Boeing E-3A AWACS and the RAF's new E-3D Sentry operate in the airborne command and control and airborne early warning roles. The RAF fleet of E-3Ds (callsign Magic) are fitted with new CFM56 engines and an air-refuelling probe. The radar has a range in excess of 300nm and can track and identify multiple aircraft at any height, as well as shipping when used in the maritime role. The controller can colour code his contacts to show helicopters, aircraft and ships etc. The aircraft is also equipped

The latest RAF Boeing E-3D Sentry provides airborne command and control for a wide range of missions. (Patrick Allen)

with the Joint Tactical Information Distribution System (JTIDS) and can down-link its information to ground users in real time. The Sentry is highly capable and is not only used for its main purpose of airborne early warning but also as airborne surveillance and command and control for COMAO multi-aircraft missions. For helicopters the Sentry is more often used as a flight following service, tracking their progress/route during a mission. There are nine mission controllers who will each be assigned to a specific task such as fighter controller, surveillance controller, link manager, electronic support measures operator etc. all watching and controlling air movements within their sectors.

USA

After the failed Operation Eagle Claw, the aborted hostage rescue mission to Tehran in 1980, the Holloway Commission was set up to review events and the commission recommended that a special operations force should be established with dedicated aircraft and crews capable of undertaking missions in all geographical regions and environments in times of peace or war at a high readiness for immediate operations. The US Army, US Air Force and US Navy would establish their own Special Forces groups and would be capable of operating independently or together under the command of the Joint Special Operations Command (JSOC) established in 1980 at Fort Bragg, North Carolina. Using existing aircraft, each service developed its own Special Operations aircraft and helicopters.

USAF Special Operations
Sikorsky MH-53J Pave Low 111
The USAF already operated a number of Sikorsky CH-53/B/Cs Pave Knifes, known as Jolly Green Giants, operating in the CSAR role, originally developed for Vietnam CSAR although

they lacked a night rescue capability. These helicopters were modified for night-time all-weather Special Operations roles and became MH-53Js, which were later further improved to the MH-53J Pave Low 111 standard. The MH-53J is capable of operating IMC down to 100 feet and below. The helicopter is equipped with inflight refuelling which extends the MH-53's 540-mile (870 km) range. The MH-53J is equipped with the AN/AAQ-10 FLIR and AN/APQ-158 Terrain Following and Terrain Avoiding (TFTA) multi-mode radar which is coupled to the autopilot and an integrated mission computer with NS/GPS/ Doppler with moving-map display, allowing the pilot to fly NOE at night at 170 knots plus at heights lower than 100 feet in visibility less than 1,000 yards. Information including FLIR imagery is displayed on CRT screens located in front of each of the pilots, with the moving map display in the centre. The helicopter is flown by a crew of six: two pilots; a flight engineer who sits in the jump seat between the pilots monitoring the instrument displays and operating the secure satellite communications and encrypted secure speech radios, plus the RWR system which includes the latest electronic and infra-red countermeasures system and the ARI18246 Nemesis (DIRCM); a second flight engineer; and two aerial gunners who man the GUA-2A/B 7.62mm and 12.7mm Miniguns. With a max. gross weight of 50,000 lb (22,675 kg) the MH-53J carries 1,000 lb (450 kg) of armour plating and can lift thirty-seven troops, sixteen litters or a 20,000 lb (9,000 kg) usable load. During 1996-7 the MH-53J Pave Low 111 fleet underwent a major cockpit upgrade to include an Interactive Defensive Avionics System/Multi-Mission Advanced Tactical Terminal (IDAS/ MATT). This system integrates all the helicopter's mission, navigation and electronics warfare systems through a MIL-STD-1553 databus to provide electronic order of battle information correlated onto a digital moving map for presentation on one of the colour MFDs. This

The Sikorsky MH-53J Pave Low Special Operations helicopters are capable of operating clandestine missions behind enemy lines in almost any weather.
(Patrick Allen)

puts all the tactical information onto a single screen which can be instantly updated in flight via satellite systems should a mobile threat appear. The system gives the pilot the big picture regarding the terrain he is flying over and how it affects threats encountered on the route; it also tells him the best height to fly at to avoid being detected, making the best use of any terrain masking. The pilot is given an electronic image of the terrain outside which alters perspective as he changes height. The system will also show the route flown, time to go, fuel burn and ETA etc. If a threat appears which requires a change in route, the system will work out the best alternative route and the speed at which to fly and still maintain time on target etc. The system not only works out routes and threats but also looks after all the defensive aids systems such as RWR, MAW and jammers etc. The system will identify a threat then launch the appropriate countermeasure and will then let the crew know what expendables remain. The system will be invaluable in terms of navigational and tactical awareness for future Special Operations missions in hostile territory, and a similar package is being installed into Special Operations HV-22B Ospreys.

Sikorsky MH-60G Pave Hawk

A variant of the Sikorsky UH-60 Black Hawk used by the USAF in the Special Operations and CSAR role, or 'Sandy' role. Used primarily for rapid SOF deployment, five Pave Hawks can be carried in a C-5A Galaxy. The MH-60G is fitted with a folding stabilator similar to the SH-60F Sea Hawk for easy folding into aircraft or below deck on ships. The MH-60G has a fully integrated avionics package with electronic displays, terrain following/avoidance radar, FLIR, extended-range fuel tanks and air-to-air refuelling probe.

MC-130 Combat Talon 11

All-weather Special Operations C-130 Hercules designed to operate at night over hostile territory undertaking infiltration/exfiltration of Special Forces and resupply missions in sensitive areas. Purpose-built for clandestine operations the MC-130 has NVG-compatible lighting, multi-function CRTs, integrated INS/GPS/Doppler navigation systems with AN/APQ 170 multi-mode terrain following/avoidance radar, FLIR and Low Level Aerial Delivery System (LLADS), allowing up to 22,000 lb (10,000 kg) of cargo to be dropped automatically at speeds of 250 knots and at heights lower than 250 feet. LLADS is automatically linked to the aircraft's integrated navigation systems using the INS and GPS to calculate all the drop parameters. The MC-130 Combat Talon 11 is the workhorse of US SOF and is regularly used to act as a FARP for the MH-53Js.

HC-130 Combat Shadow

A Special Operations C-130 variant used primarily as an air-to-air refuelling tanker for Special Operations helicopters. Designed for extended-range operations and to loiter for long periods, the HC-130 is capable of refuelling SOF helicopters at low level at night in or near hostile territory. The HC-130 has recently been updated with a full NVG-compatible cockpit, INS/GPS navigation system, FLIR and improved defensive aids suite. It has a secondary role to support MC-130 Talon 11 missions undertaking low-level aerial delivery of troops and cargo, small boats, vehicles etc. over hostile territory. The HC-130 is fitted with two 11,000 lb (5,000 kg) Benson fuel tanks, and with its own fuel tanks can carry around 82,000 lb (37,000 kg) of fuel. An MH-53J or Chinook can take on 7,000 lb (3,200 kg) of fuel in one sitting and the HC-130 can refuel two helicopters at a time from its wing-mounted drogues. Once the helicopters are confirmed inbound, the HC-130 reduces speed to around 120 knots and, without any communications between the tanker and helicopters, the tanking takes place with all crews on NVGs. There have been several missions when two or more HC-130s have flown in formation and refuelled large numbers of USAF and US Army Special Operations helicopters during large-scale combined operations.

AC-130 Spectre Gunship

Armed with two 25mm Gatling guns, one 40mm Bofors and a 105mm Howitzer all linked to an electronic and visual sensor suite which allows the AC-130 to find, identify and attack targets at night. Flying left-hand orbits the AC-130 can deliver accurate and concentrated firepower on demand from heights in excess of 6,000 feet and operates in the night close air support and interdiction role. The AC-130 can loiter over a target area for long periods and its highly sophisticated night vision/electronic sensors make the aircraft ideal as a night-time armed reconnaissance aircraft in the secondary role.

McDonnell Douglas C-17a Globemaster 111

The USAF's newest strategic and tactical airlifter is capable of operating directly from a forward operating base and under-taking tactical intra-theatre missions. The C-17A with a crew of three can carry loads in excess of 172,200 lb (78,108 kg)

A number of leased C-17 Globemaster 111s will be operated by the RAF. (Boeing)

with an unrefuelled range of 2,400nm (4,447 km) carrying a 160,000 lb (72,567 kg) payload (range is unlimited with aerial refuelling) and can include up to twelve Special Operations helicopters in its spacious cargo hold. The C-17A will supplement C-141 Starlifters and C-5 Galaxys which are used to provide the USAF and US Army Special Operations helicopters with their rapid deployment capability.

US Army Special Operations

US Army Special Operations is undertaken by the 160th Special Operations Aviation Regiment (Airborne) (SOAR) based at Fort Campbell, Kentucky. Originally formed from elements of the 101st Airborne Division (Air Assault), the 160th is equipped with dedicated Special Operations helicopters designed for night-time, long-range covert infiltration/exfiltration of Special Forces and for night attack. They include the Boeing MH-47E Chinook, Sikorsky AH/60L/MH-60L/MH60K Black Hawks and McDonnell Douglas MH-6J 'Little Birds'.

Boeing Chinook MH-47E

Originally operating modified CH-47D Chinooks, designated MH-47Ds, the 160th SOAR received the first of twenty-six MH-47E Chinooks in 1991. The MH-47E was designed to complete a five-and-a-half-hour covert mission over a 300nm (556 km) radius at low level, day or night, in adverse weather and over any type of terrain, and to do so with a 90% probability of success. The MH-47E has a fully integrated avionics system which permits global communications and navigation, and is one of the most advanced systems of its kind ever installed into an Army helicopter. The IAS includes FLIR and multimode radar (TA/TF) for nap-of-the-earth and low-level flight operations in conditions of extremely poor visibility and adverse weather. The US Army required that the Chinook MH-47E's integrated avionics system should be common and interchangeable with the US Army's Special Operations MH-60K Black Hawk. Critical components such as radios, mission computers, multifunction displays and defensive aids suites can be exchanged between the MH-47E and MH-60K allowing missions to be conducted far from the normal supply channels and securing a much higher probability of successful completion. The MH-47E has twice the fuel capacity of the standard CH-47D, an aerial refuelling system plus uprated Lycoming T55-L-714 engines with full authority digital engine management systems. The MH-47E can carry forty-four troops and Special Operations vehicles internally.

Sikorsky MH-60K

The latest Sikorsky Black-Hawk variant for US Army Special Operations has uprated engines and transmission with fully integrated avionics systems compatible with the MH-47E

A Chinook showing-off its amphibious capability which is used by Special Forces. (Boeing)

Chinook. The MH-60K is shipboard compatible, and can easily be folded to go into a ship's hangar or C-5/C-141/C-17 etc. Mission equipment such as FLIR has increased the gross weight to 24,000 lb (11,000 kg), 2,000 heavier than a normal Army UH-60L. The MH-60K is required to be capable of taking two pilots, two gunners and twelve Special Forces troops a radius of 200nm. Aerial refuelling allows unlimited range, and the MH-60K has additional internal auxiliary fuel tanks if required. The MH-60K, like the MH-47E, can be equipped with a pair of 7.62mm or 12.7mm (0.50in) Miniguns. The US Army's 160th SOAR also operates a number of Special Operations MH-60L Black Hawks fitted with FLIR, radars and IR reflective paint etc.

AH-60L Black Hawk Direct Action Penetrator

This is an armed variant of the MH-60L used by the 160th SOAR which is fitted with the Black Hawk External Stores System (ESS) and can be armed with Hellfire ASM, Stinger, M260/M261 70mm rocket pods, Mk 19 40mm grenade launcher, M230 30mm chain gun and 0.50in 12.7mm Miniguns. Equipped with TF/TA radar, FLIR and IAS, similar to the MH-47E and MH-60K, the AH-60 can operate at night far from home.

Boeing/McDonnell Douglas MH-6J 'Little Birds'

The 160th SOAR has been operating specially developed versions of the US Army Hughes OH-6 Cayuse since 1981. These small, easily transportable 'Little Birds' were developed into two variants: the AH-6, which could be armed, and a troop-carrying MH-6 Little Bird. They were the first US Army helicopters into action during Operation Urgent Fury in Grenada in 1983 when six were deployed in a C-130 Hercules. Equipped to operate by night, they have taken part in every military action undertaken by the US since 1982. They operated in the Persian Gulf in 1987 when they attacked an Iranian gunboat, took part in Panama, and were reported operating behind enemy lines during the Gulf War. The Little Birds operate almost entirely at night and recent updates have combined their armed role with the troop transport and forward air controller roles. They have now been redesignated the MH-6J. The MH-6Js are painted black with IR reflective paint, have a full defensive suite with IR jammers, satellite communications, and are equipped with NVG-compatible cockpits, FLIR, flight/mission management system, laser range-finders and a quick-change ordnance mounting system known as the 'plank' which can arm the helicopters with a variety of weapon systems including Hellfire, Stinger, Hydra 70mm rockets, 0.5in machine-gun and M134 7.62mm Minigun. The 'plank' can also be used to carry four Special Forces troops outside the helicopter (six in total). The MH-6J has a 400nm range and Little Birds can be unloaded from a C-130 and be flying in fewer than ten minutes. Twenty-two of

A US Army 160 SOAR Special Operations MH-6 'Little Bird'. (Mike Coombes)

the helicopters can be carried in a C-5 Galaxy or twelve in the new C-17 Globemaster. Special Operations Little Birds will continue operating into the next century and are slated for a mid-life extension upgrade which will include new six-blade tail rotor and more powerful Allison 250-C30R-3 gas turbine engines developed for the US Army's OH-58D armed Scout; they will also be fitted with Full Authority Digital Engine Control (FADEC). Role equipment upgrades will include improved night vision, communications and navigation systems, a lightweight universal 'plank' for weapons, fuel tanks and external personnel systems including lightweight Hellfire anti-armour missiles, integrated weapons management and laser designating systems.

US Navy Special Operations
Sikorsky HH-60H

The US Navy operates a modified version of the SH-60F Sea Hawk in the Special Operations role, designated the HH-60H. The helicopter has two primary missions, CSAR and Special Warfare Support (SWS) for the US Navy SEALs. Configured in the SWS role, the HH-60H can deploy eight SEALs 237nm, or over a radius of 189 miles. With a crew of four, the HH-60H has APR-39 radar warning receivers, ALQ-144 infra-red jammers, engine exhaust infra-red suppressors, NVG-compatible cockpit and armoured crew seats. The helicopter has ALE-39 chaff/flare dispensers and can be armed with M60Ds or the Minigun. The HH-60H is operated by HCS-5 (Pt

Magu, California) and HCS-4 (Norfolk, Virginia). The United States Marine Corps undertakes Special Operations but uses its existing helicopters and aircraft along with Special Operations capable aircrews. Helicopter squadrons train as part of the Marine Expeditionary Unit (Special Operations capable) using existing helicopters such as the Sikorsky CH-53E, Boeing CH-46E, Bell AH-1W Cobras and Bell UH-1N Hueys undertaking long-range insertions of US reconnaissance marines and SEALs. All Special Operations capable marine aviators are NVG-qualified and regularly practise hostage rescue, CSAR/TRAP and counter-terrorist plus long-range deep strike missions from the sea.

Future Special Operations Aircraft
Bell Boeing V-22 Osprey

The USAF Special Operations Force Command has ordered fifty CV-22B Special Operations tilt-rotor transports, the first of which will be delivered in 2003, and these will eventually replace USAF SOF MH-53Js, MH-60Gs and a number of MC-130 and HC-130s. The CV-22B can take off and land vertically, but unlike a helicopter it can cruise at 275 knots and has a range in excess of 515nm, which can be extended with a short take-off run. With aerial refuelling range is unlimited and restricted only by crew fatigue. Many of today's Special Operations helicopters fly with two crews, swapping over during a mission. The CV-22B can carry twenty-five Special Forces troops and is capable of world-wide deployment. With

The Bell/Boeing MV-22B Osprey enters service with the USMC in 1999 and will replace a number of Special Operations helicopters and aircraft. (Boeing)

An artist's impression of the Avpro exint pod fitted to a Sea Harrier. (Avpro/via David Oliver)

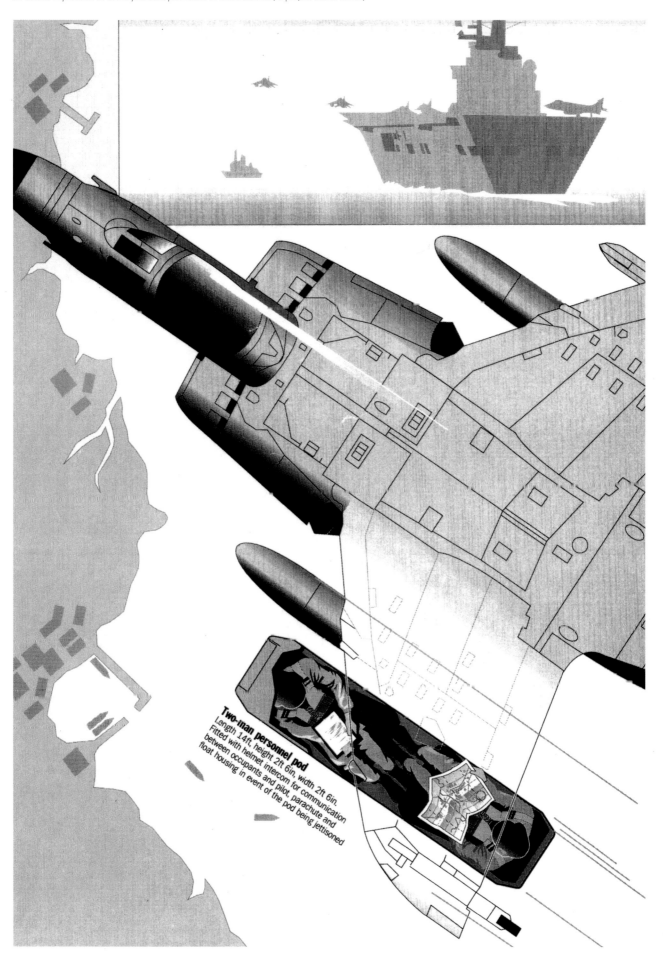

Two-man personnel pod
Length 14ft, height 2ft 6in, width 2ft 6in.
Fitted with helmet intercom for communication
between occupants and pilot, parachute and
float housing in event of the pod being jettisoned

extended-range tanks it can fly over 1,800nm without the need for an aerial refuel. Marine Corps MV-22Bs are destined for combat assault and assault support, the US Navy's HV-22Bs for the Combat SAR and Special Warfare (SEAL) support. All V-22s are shipboard compatible and can be folded down for below-decks storage. Powered by twin Allison Gas Turbine T406-AD-400 powerplants the CV-22B has onboard computer management systems that are designed to reduce pilot workload and will include the IDAS suite fitted to the MH-53J Pave Lows. Equipped with a fly-by-wire flight control system the CV22B has a fully computerised flight management system and avionics package which monitors all vital systems and instantly shows all flight, navigation and tactical data on large full-colour video displays (CRTs). The CV22B will be operated for much of the time in the aeroplane mode, which will reduce fuel burn using much less fuel than a helicopter and reduce the need for tanker support, while also extending range and reducing mission times.

Boeing Sikorsky RAH-66 Comanche

The RAH-66 Comanche has been designed as a low observable helicopter presenting the smallest possible cross-section for radar, infra-red and acoustic signatures. As an armed reconnaissance, light attack and air combat capable helicopter, the RAH-66 is ideally suited to undertaking long-range deep strike or armed escort support for Special Operations. Built of fibre-reinforced composites and powered by the twin T800 engines plus a five-bladed bearingless main rotor and 'FANTAIL' anti-torque system, the RAH-66 is fast (175 knots) and is capable of day/night poor weather operations. Mission equipment includes centralised processing, triple redundant fly-by-wire flight control systems, self-healing digital mission electronics, triple redundant electrical and hydraulic systems, low workload crew station with two six-by-eight-inch multifunction displays, passive long-range high-resolution sensors, wide-field-of-view helmet-mounted displays, simple remove and replace maintenance, fully retractable missile armament system and stowable three-barrel 20mm Gatling gun. In the armed reconnaissance role the RAH-66 can be armed with four AGM-114 Hellfires plus two AIM-92 Stingers and 320 rounds for the Gatling gun. Fully loaded for a fight the RAH-66 can carry a maximum of fourteen AGM-114 Hellfires or twenty-eight AIM Stingers, or fifty-six 70mm (2.75in) rockets.

A full night capability is attained by using passive second-generation sensors and infra-red viewing systems that will look wherever the pilot looks using the helmet-mounted sighting system, or which can be displayed on the cockpit CRTs, identical for both crew members. RAH-66 can self-deploy 1,260nm and with digital communications and tactical displays is compatible with the US Army's future digitised battlefield network, and would prove a useful addition to the US Army's Special Operations Force replacing the MH-6Js in the escort and attack roles.

Avpro Ltd, wing-in-ground-effect air vehicles

The British company Avpro Limited has been designing future Special Operations air vehicles and man-carrying pods which have shown potential for future operations.

Avpro exint pod

Exint (extraction/insertion) is a two-man pod which can be fitted onto the weapons pylon of an attack helicopter like the WAH-64D Apache or RAF/RN/USMC Harriers and can be used to insert/extract Special Forces, for CSAR of downed aircraft, casevac from remote areas and ships and emergency deployment of key personnel and equipment. The fourteen-foot (4.27m) long pod can carry two people or the equivalent weight in equipment and is fitted with helmet intercoms for communication with the pilot, and a parachute and floating housing in the event of being jettisoned in an emergency. It will also be fitted with Avpro's Satellite-Assisted Recovery System (SARS).

Avpro Marauder and Manta

The Russian Navy spent many years experimenting with wing-in-ground-effect vehicles obtaining very high speeds. Avpro Limited has expanded on this concept to propose its two new Assisted Hull Lifter Vehicles (AHLVs). The Marauder and the Manta both use jet engines to pump exhaust air at high pressure under wings, giving them lift and allowing the vehicles to operate at high speeds just above the surface of the water, riding on a cushion of air. The Marauder is being proposed as a small patrol craft capable of anti-submarine/shipping roles and the insertion/extraction of Special Forces/Marines etc. The larger Manta is to operate from assault ships and carry cargo and loads in excess of forty tonnes at speeds of over fifty knots.